Proboscis Monkey

The Natural History of the
Proboscis Monkey

John C.M. Sha, Ikki Matsuda and Henry Bernard

Natural History Publications (Borneo)
Kota Kinabalu

2011

Published by

Natural History Publications (Borneo) Sdn. Bhd. (216807-X)
A913, 9th Floor, Wisma Merdeka Phase 1,
P. O. Box 15566,
88864 Kota Kinabalu, Sabah, Malaysia.
Tel: +6088 233098 Fax: +6088 240768
e-mail: info@nhpborneo.com Website: www.nhpborneo.com

The Natural History of the Proboscis Monkey
by John C.M. Sha, Ikki Matsuda and Henry Bernard

ISBN 978-983-812-130-9

First published January 2011

Half-title page: A group of proboscis monkeys sleeping on a giant fig tree (*Ficus benjamina*) at dusk by the Kinabatangan river in Sukau, Sabah. Photo: Ch'ien C. Lee.

Design and layout by Chan Hin Ching.

Printed in Malaysia.

CONTENTS

Foreword

Worldwide, primates are one of the most vulnerable groups in the animal kingdom, particularly over the last 50 years when habitat destruction and excessive hunting in many areas has caused a serious decline in populations around the globe. This is also the case for the unique proboscis monkey, which is endemic to Borneo, and "Endangered", as classified by the International Union for Conservation of Nature (IUCN).

The proboscis monkey has not received the attention that it deserves from the conservation community as well as the general public. Some of the contributing reasons are the lack of information, research and longer-term monitoring efforts that can be applied to sound conservation and management decisions, as well as wider awareness of the conservation plight of the species.

In this light 'The Natural History of the Proboscis Monkey' is a timely publication and particularly relevant as, apart from presenting novel research data on the species, it also highlights some of the areas where urgent action is needed to safeguard the long-term survival of proboscis monkeys. This includes translocation of highly isolated populations, and re-establishment of forest and river corridors to compensate for the many fragmented populations of proboscis monkeys; and it recommends specific research priorities going forward, including long-term monitoring of population numbers and distribution, and more accurate habitat and land use classifications.

'The Natural History of the Proboscis Monkey' also attempts to reach out to a wider audience, by presenting information that the general public or wildlife enthusiast may be interested in. From an eco-tourism point of view, the importance of the proboscis monkey as a flagship species cannot be underestimated as unlike many other charismatic wildlife species, such as the Bornean Orang Utan and Bornean Elephant, the proboscis monkey is an easily observable animal with very high encounter rates in the wild. Hence, contained in this book is useful information of recommended places to visit to observe the proboscis monkey as well as recommended guidelines for observing them so as to minimise disturbance and to promote sustainable tourism.

Wildlife Reserves Singapore is proud to have the largest primate collection in the world with an outstanding breeding record. This includes most significantly, the proboscis monkey which is notoriously difficult to maintain and breed in captivity. In the last 12 years, we have had 19 successful births from our captive breeding programme and a collection that constitutes the largest breeding group outside its range countries. Captive-based research has contributed to important knowledge on proboscis monkeys that could not be obtained from wildlife research, like aspects of diet, disease and life history information. Therefore it is important to establish a constructive *in-situ* and *ex-situ* collaborative link to provide the necessary impetus for conservation and education initiatives, which will be beneficial to the long-term conservation of the proboscis monkeys.

It has been a privilege for Wildlife Reserves Singapore to have a opportunity to support this important effort, and I would like to congratulate the authors for completing '*The Natural History of the Proboscis Monkey*', which I am sure will inspire others to follow this pioneering effort.

Fannyle.

Ms. Fanny Lai
Group CEO, Wildlife Reserves Singapore

INTRODUCTION

The machinations of fate have converged our paths upon the tip of the third largest island in the world — the state of Sabah on Malaysian Borneo. A mysterious place seemingly untouched by the age of globalisation and, for the uninitiated, one which can still conjure up wild images of impenetrable rainforests, impassable mountain ranges, undiscovered head-hunting tribes and a diversity of weird and untamed animals beyond one's most vivid imagination.

Travelling along the water highways of the major rivers and estuaries in Sabah, one is bound to be drawn to the unmistakable sights of little orange visages breaking through the monotonous green and brown of the forest fabric, and the boisterous "honks" and "growls" awakening the tranquillity of the forest silence. This is what brought us together on this common journey of discovery, an animal that astounded early explorers and which still fascinates many today — one

that is "amazingly wonderful", yet "highly ludicrous" and "amazingly grotesque" — the proboscis monkey.

The proboscis monkey is an endangered primate species found only on the island of Borneo. They live mainly in mangrove, swamp and riverine habitats along waterways. Proboscis monkeys are increasingly threatened throughout their range due to their habitat specificity and specialised diet which is a bane to their survival as human activities like expansion of human habitation and forest clearance for timber, large-scale agriculture and other purposes have become particularly intense in these areas, resulting in fragmented and disturbed populations in forest "Alcatrazes". The rivers and freshwater swamps of Sabah represent one of the last vestiges of natural habitat for proboscis monkeys, where viable populations still thrive.

Above: The mysterious forests of Sabah evoke vivid images of weird and untamed animals.

While the distant cousin of the proboscis monkey, the orangutan, has received much attention from the international scientific community and the general public in addressing its conservation plight, the proboscis monkeys has received comparatively little notice despite increasing evidence of realistic threats to its long-term survival. Contributing to this is the fact that large numbers are easily sighted along waterways where they congregate to rest, giving the impression that they are abundant and unthreatened, as well as the dearth of detailed information about these monkeys due to the difficulties associated with studying them in their swampy habitats. Through the many interactions with our nonproboscis-specialist colleagues and the general public, we recognise that many misconceptions and knowledge gaps about proboscis monkeys still exist, for example, the proboscis monkey is commonly thought of as a pure mangrove species, which is not true as its habitat is more varied and actually includes freshwater

The unique features of t he proboscis monkey astounded many early explorers and still fa scinate many today.

and peat swamps as well as dipterocarp forests. Our recognition of such misunderstandings proved that it is timely to address these issues, as well as to re-emphasise the urgent need for conservation action for this species.

Based on scientific research conducted by the authors and additional insights from the work of numerous other researchers, *The Natural History of the Proboscis Monkey* presents an updated account of the natural history of proboscis monkeys with particular focus on Sabah where a wealth of research information from the authors, currently privy only to the scientific community, remains to be translated for the benefit of a wider audience and contribute to the longer-term goals of conservation education.

In this book, we highlight novel findings on proboscis monkey ecology and behaviour in the last fifteen years, new threats that have not been sufficiently considered previously, and put this information

Proboscis monkeys live in habitats along waterways that are increasingly becoming threatened.

in the context of the interconnectedness of proboscis monkeys with their unique habitats and human communities along Sabah's rivers of life. Chapter 2: "Sabah: Its forests, Diversity and People" presents a background of the natural history of Sabah with emphasis on the habitats, wildlife and people that are closely associated with proboscis monkeys. Chapter 3: "Studying the Proboscis Monkey" discusses the difficulties associated with studying proboscis monkeys and introduces the knowledge gaps that we attempt to address in the rest of the book. Chapter 4: "Proboscis Monkeys and Their Habitats" introduces the life history of proboscis monkeys and the habitats they are found in. Chapter 5: "Distribution of Proboscis Monkeys" presents the most up-to-date information on the distribution and population status of the proboscis monkey in Sabah, as well as important areas for proboscis monkeys throughout its range in Borneo. Chapter 6: "Activity and Behaviour" illustrates the typical daily activities and unique behaviours of proboscis monkeys. Chapter 7: "Social Organisation" describes their social structure and discusses novel findings on the social organisation of proboscis monkeys. Chapter 8: "Feeding Ecology" highlights the unique foraging and dietary diversity and adaptability of proboscis monkeys. Chapter 9: "Ranging Behaviour" looks at the home range and daily ranging patterns of proboscis monkeys. Chapter 10: "Natural Predation" discusses the natural predators of proboscis monkeys and the effects on their river crossing behaviour and sleeping site selections. Chapter 11: "Threats to Proboscis Monkeys" highlights the immediate and longer-term threats that proboscis monkeys are facing, based on evidence from recent studies. Chapter 12: "Conservation of Proboscis Monkeys" discusses conservation measures needed to safeguard the survival of proboscis monkeys. Chapter 13: "Ecotourism" provides information on places to visit to view proboscis monkeys in Sabah and observation guidelines to minimise disturbance and human impacts on their natural behaviour.

Opposite: Adult male proboscis monkey, a charismatic animal. Many misconceptions and knowledge gaps still exist for the proboscis monkey.

Aerial view of the rainforest canopy.

02

SABAH:
ITS FORESTS, DIVERSITY
AND PEOPLE

Sunset view of Mount Kinabalu, the peak of which is the highest point in Southeast Asia.

The island of Borneo in Southeast Asia is the world's third largest island and is widely considered an important centre for biological diversity. The Malaysian state of Sabah, with a land area of approximately 7.36 million hectares, is situated at the northern tip of Borneo, with Malaysian Sarawak, Indonesian Kalimantan and Brunei Darussalam making up the rest of the island.

The landform of Sabah is predominantly hilly. The western and central regions of Sabah are dominated by rugged hill and mountain ranges with elevations from 300 to 1200 metres above sea level. The Crocker, Trus Madi and Maligan ranges exceed 1500 metres elevation at many points and run almost parallel to the west coast, extending from the southern end of Marudu Bay in the north southwards along the western part of the state to the Sarawak border. Standing at a height of 4095 metres, Mount Kinabalu, situated at the northern limits of the Crocker Range, is the highest mountain in Southeast Asia. The narrow western lowland plains contain areas of low, flat ground and include a number of off-shore islands. These lowlands are rather densely inhabited. Eastern Sabah is characterised by low dissected hills, gentle slopes and poorly drained flatland and low-lying swampy zones embracing the deltas. Tracts of gentle slopes are scattered and rarely extensive, while the most extensive areas of flat land are subject to waterlogging or floods.

A wooden house with roof made from the fronds of nipah palms located in Abai, Kinabatangan, Sabah.

Sabah's climate is marine equatorial, with high rainfall (about 2500–3500 mm annually). Typical annual rainfall distribution patterns vary in different parts of Sabah due to the influence of coastal and shadowed land-masses or ranges. In western Sabah, rainfall is usually concentrated between May and June during the south-west monsoon season and between October and November, at the beginning of the north-east monsoon season. In north-eastern Sabah, there is usually heavy rainfall throughout the north-east monsoon period from November to February. The rainfall patterns tend to be more evenly distributed throughout the year in inland Sabah compared to areas nearer the coast. Long periods of drought or dry spells with very little

Dipterocarp forest.

Coastal mangrove forest.

Nipah forest.

Swamp forest.

Riverine forest.

Logged forest.

Banana plantation.

Oil palm plantation.

(less than 500 mm for six months) or no rain also occur frequently. Although there are occasional severe storms in the coastal areas, the entire state of Sabah lies outside the typhoon belt. Relative humidity on the lowlands usually exceeds 80% with night and day temperatures ranging from 20°C to 34°C.

Forests of Sabah

Dipterocarp forest is the major natural forest type found in inland Sabah: Lowland dipterocarp forest occurring at less than 300 metres

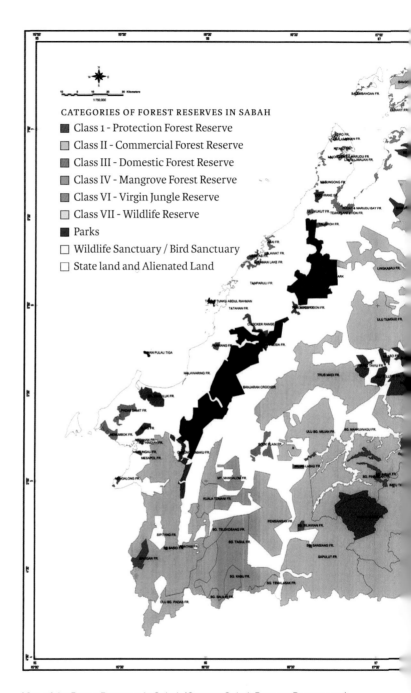

Map of the Forest Reserves in Sabah (Source: Sabah Forestry Department).

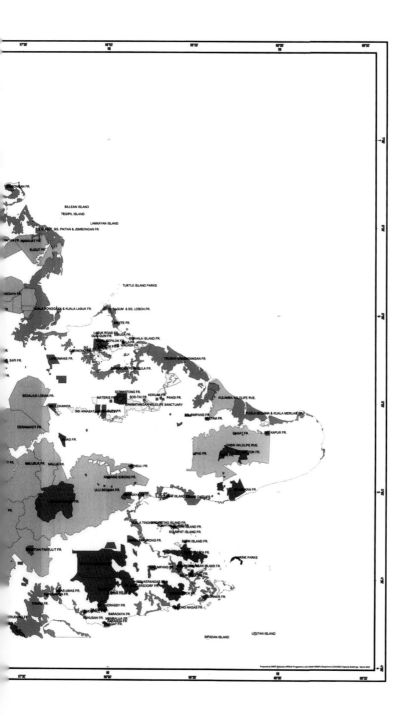

BILLEAN ISLAND

TEGIPIL ISLAND

LANKAYAN ISLAND

TURTLE ISLAND PARKS

SABAH: ITS FORESTS, DIVERSITY AND PEOPLE   15

above sea level, Hill dipterocarp forest at less than 750 metres, Highland dipterocarp forest at less than 1200 metres above sea level. Above these elevations, Montane, Cloud and Sub-alpine forest are found. Nearer the coast and river deltas, swamp forests (peat swamp and freshwater swamp), mangrove and nipah forest and riverine forest (dipterocarp forests along rivers which are distinct from those in the hinterland in terms of species composition) are found.

Like most parts of Borneo, human activities have had a considerable impact on Sabah's vegetation. The original forest cover of Sabah would have been an almost complete cover of natural forests until it was initially depleted by logging, tobacco and rubber plantations with the advent of colonial rule in the late 19th century. The trend of forest conversion continued in subsequent years with the inevitable increase in agricultural crop cultivation, logging and expansion of human habitation. The western and northern regions of Sabah have mostly been developed for human habitation with expansion in Sabah's human population, while the eastern regions have been extensively logged and converted to permanent agriculture. Current total natural forested area in Sabah is estimated to be 4.1 million hectares or 55% of total land area.

Forest Reserves in Sabah

The Sabah State Government has established many forest conservation areas so that the rich natural heritage may be protected by law. Forest reserves in Sabah total 3.61 million hectares or 48% of the total land area and are classified into seven classes according to the level of protection and usage of the forest and resources. Class II Commercial Forest, Class III Domestic Forest, Class IV Amenity Forest, and Class V Mangrove consist of various forests allocated for logging, consumption by local communities, provision of amenities, and recreation for local inhabitants, as well as to guarantee a supply of mangrove, timber and other produce to meet general trade demands. Class I Protection Forest, Class VI Virgin Jungle Forest and Class VII Wildlife Reserve consist of protected forests conserved for the protection of watersheds and the maintenance of the stability of essential climatic and other environmental factors, as well as research on and protection of wildlife. The Wildlife Conservation Enactment 1997 provision for three additional types of protected areas: Conservation Areas for fast and flexible protection of wildlife and habitats; Wildlife Hunting Areas for animal population management by regulated hunting; and Wildlife Sanctuaries, the strongest conservation category for fauna, flora, genetic resources and habitats. The Sabah Parks Enactment 1984 also regulates the use of terrestrial and marine parks. Some of the important protected conservation areas and parks include Tabin Wildlife Reserve, Kinabatangan Wildlife Sanctuary, Crocker Range Park, Kinabalu Park, and the Danum Valley and Maliau Basin Protection Forest Reserves.

Diversity of Life in the Floodplains

Sabah features many swampy, forest-covered floodplains which can support particularly rich ecosystems, both in quantity and diversity. The floodplains of Sabah are home to a myriad of species. Within the Kinabatangan floodplains alone, 250 bird species, 90 fish species, 20 reptile species, 50 mammal species and 1056 plant species have been recorded.

This variety of wildlife in the floodplains includes birds like the rhinoceros hornbill (*Buceros rhinoceros*), blue-eared kingfisher (*Alcedo meninting*), grey-headed fish-eagle (*Ichthyophaga ichthyaetus*) and oriental darter (*Anhinga melanogaster*), Chinese egret (*Egretta eulophotes*) and cattle egret (*Bubulcus ibis*); mammal species like the Bornean Pygmy elephant (*Elephas maximus borneensis*), bearded pigs (*Sus barbatus*), sambar deer (*Rusa unicolor*), common palm civet (*Paradoxurus hermaphroditus*), Malay civet (*Viverra tangalunga*), Flying Fox (*Pteropus vampyrus*), Spotted-winged fruit bat (*Balionycteris maculata*), clouded leopard (*Neofelis diardi*), marbled cat (*Pardofelis*

Bearded pig (*Sus barbatus*).

Slow loris
(*Nycticebus coucang*).

Spotted-winged fruit bat
(*Balionycteris maculata*).

Bornean Pygmy elephant
(*Elephas maximus borneensis*).

Malay civet or Tangalung
(*Viverra tangalunga*).

Bornean Orangutan
(*Pongo pygmaeus*).

marmorata), and leopard cat (*Prionailurus bengalensis borneoensis*). Ten of the thirteen primate species found on Borneo are found in Sabah and on these floodplains. These include the Bornean orangutan (*Pongo pygmaeus*), Bornean gibbon (*Hylobates muelleri*), long-tailed macaque (*Macaca fascicularis*), pig-tailed macaque (*Macaca nemestrina*), silver langur (*Trachypithecus cristatus*), maroon langur (*Presbytis rubicunda rubicunda* and *P.r. chrysea*), Hose's langur (*Presbytis hosei*), Slow loris (*Nycticebus coucang*) and Western tarsier (*Tarsius bancanus*). These floodplains are also home to the proboscis monkey.

The Orang Sungai village in Kampong Sukau, Kinabatangan.

The People and Community

Sabah is a cultural melting pot of ethnic diversity with indigenous communities consisting of some thirty-nine ethnic groups speaking more than fifty languages and eighty dialects. Some of these indigenous communities live mostly in the rural areas where about 70 per cent of Sabah's population still resides. The diverse landform, soils, climate and vegetation provide a diverse agro-ecosystem for the many indigenous communities, who are mostly subsistence farmers, utilising traditional farming practices. Amongst the indigenous groups, the Dusunic, Murutic and Paitanic groups are the largest, each occupying distinct geographical areas of settlement.

The Paitanic group reside mainly in the north-eastern part of Sabah, along the major eastern rivers that feed Sabah's east coast, such as the Paitan, Sugut, Kinabatangan and Segama rivers. Many peoples of the Paitanic group are Muslims (converted by Islamic missionaries who arrived in the 15th century) and they call themselves the Orang Sungai or River People. Historically, they built their homes on the floodplains where water and fertile land are abundant. River transportation was a key economic factor in the founding of many such communities. Apart from swidden agriculture and subsistence farming, many of the land-based indigenous communities rely on the diverse plants in the forest for their food, medicine, fuel, building materials and other household needs. Along the coastline and river mouths, there are many fishing communities. Their cash income is derived from surplus food crops, cash crops, jungle produce and fish sold in the market. Trade brought about by the Chinese (perhaps dating as far back as 2,000 years) also contributed to the economy of these river communities. Sabah's jungles were famous for products which were highly esteemed: rattans and bamboos, feathers for fashion, bezoar-stones and deer horn for medicines, rhinoceros horn as an alleged aphrodisiac, the casque of the huge Hornbill birds (prized as 'golden jade' by Chinese carvers), camphor and other woods, honeybees, wax, sago.

The close dependency of the river people on nature, their religious beliefs as well as a traditional culture of respect for wildlife that is shared by most of the indigenous groups in Sabah means a generally peaceful coexistence between people and wild animals along these rivers of life. Incidentally, the floodplain homes of the Orang Sungai also coincide with the major concentration of proboscis monkeys found in Sabah today.

Boats are still the primary means of transportation for Orang Sungai.

03

STUDYING THE
PROBOSCIS MONKEYS

Although substantial research has been conducted on proboscis monkeys, our knowledge about them is still incomplete. This is mainly attributed to difficulties in observing and tracking proboscis monkeys for long durations because of the inhospitable habitats they are found in. The relatively shy nature of proboscis monkeys (they tend to flee from observers upon visual or auditory contact) also presents a genuine challenge for researchers to get close enough to these monkeys to observe their behaviours in detail.

Our knowledge about proboscis monkeys so far is drawn mainly from independent studies in a few localities and habitats of known populations, for example, Brunei Bay in Brunei, Samunsam Wildlife Sanctuary in Sarawak, Tanjung Puting National Park in Kalimantan and Kinabatangan River in Sabah. Many of the early research studies were conducted solely in mangrove habitats using boats, which present numerous limitations such as low tide or rivers blocked by fallen trees and inhibiting attempts to track the monkeys for extended durations. Studies conducted by boat also limit our understanding of proboscis

Studying the proboscis monkey in inland forests is important to complement boat-based studies.

(Left). On land, access is difficult at places where proboscis monkeys are normally found. (Right). Due to the swampy habitats of the proboscis monkey, most studies are conducted using boats.

monkey behaviour in inland forests; since proboscis monkeys typically return to riverbank areas for sleeping, and travel inland away from the riverbank from early morning to late evening, thus limiting the number of observation hours on their full diurnal activity if only surveys on boats are used. Although radio telemetry has been attempted in one study, the sensitivities of disrupting social group structure in the process of sedating an animal to attach necessary equipment means this method can only be applied on rescued or captured monkeys. Due to these difficulties, despite numerous behavioural and ecological studies on proboscis monkeys to date, large knowledge gaps about proboscis monkeys still exist, for example:

1. There is no accurate basic information about proboscis monkey distribution and population status through its range, which impedes more detailed population viability analysis for the formulation of conservation action plans;

2. The inter-group social organisation of proboscis monkeys and associations between groups away from the riverbanks and in inland forests are still unclear;

3. Knowledge about the foraging ecology of proboscis monkeys is incomplete because conclusions are based mainly on observations at sleeping sites;

4. Predation threats have not been extensively examined as only waterborne predators close to sleeping sites have been considered;

5. Threats to proboscis monkey populations across their range have not been adequately examined.

Through large-scale studies conducted across the entire state of Sabah and detailed examination of selected groups at long-term study sites in Klias Peninsula and Kinabatangan, including the habituation of groups for full-day tracking by foot in inland forests, we were able to get a more in-depth understanding of proboscis monkeys and to address some of the above-mentioned knowledge gaps on the species.

Boat studies are limited by fluctuating tide timings and accessibility.

04

PROBOSCIS MONKEYS
AND THEIR HABITATS

Adult males (left) have large pendulous nose compared to pointy witch-like in females (right).

The proboscis monkey (*Nasalis larvatus*) is classified under the family Cercopithecidae or Old World Monkeys — one of the largest and most diverse primate families distributed throughout Africa and Asia; and subfamily Colobinae, representing the leaf-eating monkeys. Colobines are medium-sized primates with long tails and are almost exclusively herbivores, feeding mainly on leaves, flowers and fruits. They have a complex stomach for bacterial and enzymatic digestion of these hard-to-digest plant materials. Proboscis monkeys have reddish-orange fur on the crown and back. The shoulders, cheeks, throat, and nape are pale orange. The proboscis monkey has partially webbed feet, but the front and hind limbs are similar in length, which has been suggested to be characteristic of more terrestrial primates. In fact, there are indications that the genus *Nasalis* was a predominantly terrestrial monkey reminiscent of macaques, evolved in forest woodland characterised by an openness demanding considerably more terrestrial anatomical adaptations. Prior to 13 million years ago, Colobinae and Ceropithecinae shared a common ancestor *Victoriapithecus*, which did not have a specific stomach adapted to digesting leaves and had a relatively high tendency to feed on fruits. The split between African and Eurasian colobines then occurred approximately 12 million years ago, followed by the divergence among Asian Colobinae some 3 million years ago to the proboscis monkeys we see today.

Life History

Life history information on proboscis monkeys is mostly obtained from captive animals as little information on these aspects is available from the wild. Proboscis monkeys have an average lifespan of 13.5 years. The oldest surviving proboscis monkey at Singapore Zoo, a female, is 17 years old and is still reproductively healthy. Proboscis monkey infants are typically up to 12 months old, but in the wild they are dependent on their mother for up to 24 months. Infants are born with a dark blue face which slowly turns greyish and then pink, like the adults. Juveniles are from 12 to 36 months. Males reach sexual maturity at 60–84 months while females reach maturity at 36–60 months. Inter-birth interval is 12–24 months and gestation period is 166 days.

Male or Female?

Proboscis monkeys are sexually dimorphic. Adult males (~21.2 kg) weigh on average twice as much as females (~10 kg). Male body length averages 74.5 cm (range 73–76 cm) compared to female body length of 62 cm (range 55–62 cm). Tail length of the male averages 66.5 cm (range 66–67 cm) compared to the female average of 57.3 cm (range 55–62 cm). On sight, mature proboscis monkey males are easily distinguishable from females by their large elongated pendulous noses giving the proboscis monkey its common name — "Monyet Belanda" (Dutch monkey). Only the dominant males have the large noses; females have sharp pointy witch-like noses. The

Infants less than a year old have a greyish-blue face (Top L) which slowly turn grayish (Top R) and then pink like the adults. The rump patch is white in adult males (Bottom L) and greyish in females (Bottom R).

legs, belly, rump patch, and tail are whitish grey in females and white in males. The striking red male penis and black scrotum also make the male easily distinguishable at close range.

Proboscis monkeys have partially webbed feet that allow them to walk on soft mangrove floor without sinking, as well as for swimming.

Proboscis monkeys use both their hind and fore limbs for horizontal locomotion.

Proboscis Monkey Habitats

Early work on the proboscis monkeys suggested that they are dependent primarily on mangrove forests for food and cover, but more recent studies showed that proboscis monkeys also use extensive riverine and other swamp habitats. In Sabah, proboscis monkeys are known to use three main forest types — riverine forest, mangrove forest and swamp forests (peat swamp, freshwater swamp). In total, mangrove, freshwater swamp, and undisturbed dipterocarp forests account for only about 9.8% of the total land area in Sabah.

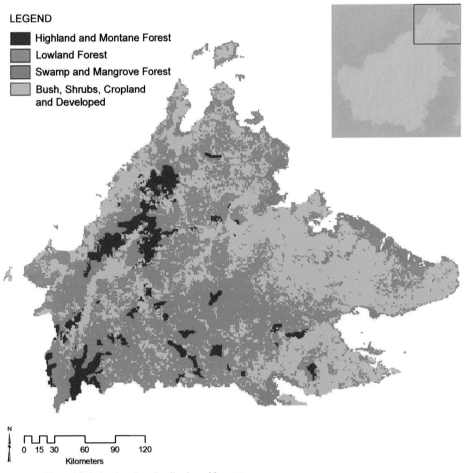

LEGEND

■ Highland and Montane Forest
□ Lowland Forest
□ Swamp and Mangrove Forest
□ Bush, Shrubs, Cropland and Developed

N

0 15 30   60   90   120
Kilometers

Map of Sabah, showing distribution of forest types.

Proboscis monkeys on the mangrove floor at low tide.

Habitat preference

In Sabah, proboscis monkeys are found most abundantly in riverine forests and mangrove forest. Densities are comparatively much higher in riverine forest than in mangrove forest, even at sites where continuous riverine and mangrove habitats are present. Although densities can be even higher in swamp forests, these habitats represent only a small proportion of proboscis monkey habitats in Sabah compared to riverine and mangrove forests. Proboscis monkeys are seldom found in extensive stands of nipah forest, but habitat mosaics of mangrove and nipah can support high densities of proboscis monkeys. Studies have shown that preferred habitats may be important to the proboscis monkey during certain periods of the year, with seasonal migrations between habitats observed at Sukau in Sabah and Bako National Park in Sarawak.

Non-proboscis Habitats

Proboscis monkeys have been recorded in disturbed habitats of secondary growth near human settlements, in remnant tidal forest close to agricultural land and in selectively felled forest, but they are not known to use many habitats, in particular farmland and permanent cultivations such as oil palm plantations. They are also known to occur only at less than 350 metres above sea level and do not use highland or montane forests.

Adult male proboscis monkey sitting on a mangrove tree.

05

DISTRIBUTION OF
PROBOSCIS MONKEYS

The Asian Colobinae range from 35°N in Pakistan to 9°S in the Malay Archipelago, a geographic distribution approximating that of the Oriental zoogeographic region to Borneo, where they are found in the Malaysian states of Sabah and Sarawak, Indonesian Kalimantan and Brunei Darussalam. The restricted and disjunct distribution of proboscis monkeys appears unusual. A biogeographic analysis of the extant Asian Colobinae indicated that at about 190,000 years Before Present, a major deforestation exterminated *Nasalis* in Sumatra with the termination of the interglacial period, leaving the Bornean proboscis monkey separated from Sumatra. Its island distribution prevented it from following the climatic and geographic recession of its native vegetation when intermittent colonisation by mangrove and lowland rainforest occurred during glacial periods. It is postulated that the swimming ability of proboscis monkeys reduced its dispersal ability through rafting because it would have had less reluctance about deserting a raft and a greater ability to resist wind and currents.

Proboscis monkeys are arboreal primates that spend most of their time in the trees but, given their exceptional ability to swim, their restricted distribution appears unusual.

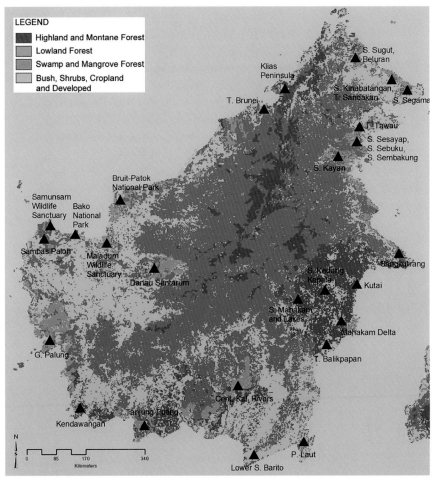

Major populations and priority areas for proboscis monkeys in Borneo (from Bennett & Gombek 1993, Meijaard & Nijan 2000, and Sha *et al.* 2008).

Population and Distribution of Proboscis Monkeys in Sabah

It is estimated that 6,000 proboscis monkeys are still found in Sabah, along most coastal river systems where suitable habitats still exist. On the west coast, populations of proboscis monkeys were found in the Klias Peninsula (five sub-populations of 818 individuals in 75 groups) [1–5]. On the east coast, populations were found at Tangkarason and

Paitan (90 individuals in eight groups) [6,7]; Sugut River (787 individuals in 58 groups) [8]; Beluran (317 in 30 groups) [9]; Sandakan (three sub-populations of 326 in 28 groups) [10–12]; Kinabatangan River (1,454 individuals in 101 groups) [13]; Segama River (1,040 individuals in 83 groups) [14]; Lahad Datu (four sub-populations of 188 individuals in 16 groups) [15–18]; Semporna Peninsula (four sub-populations of 169 individuals in 16 groups) [19–22]; and Tawau Bay (718 individuals in 63

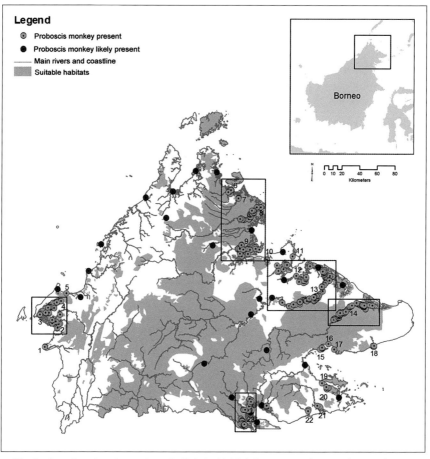

Distribution of proboscis monkeys in Sabah, highlighting areas where major continuous populations exist.

groups) [23]. Proboscis monkeys may still be present in several locations on the west coast — Bongawan, Tempurong, Pulau Gaya, Rampayan; north coast — Pitas, Marudu Bay; and east coast — Bongaya, Labuk Bay, Gum Gum, Sekong Bay, Mumiang, Lokan, Dewhurst Bay, Kulamba Forest Reserve, Tinkayu, Silam, Pulau Sebatik and Kalabakan.

Proboscis monkeys in Sabah, like findings in Kalimantan, are found most abundantly at distances less than 50 kilometres from the coast. The range limits of proboscis monkeys in Sabah are likely to extend much further inland, as far as Danum Valley *c.* 170 kilometres and Maliau Basin c. 200 kilometres. There are also recent inland records in the vicinity of Serinsim near Marak Parak in Kota Marudu and in Ulu Tungud Forest Reserve and Dermakot Forest Reserve. Other inland records from literature showed that remnant populations are still present and may represent the actual distribution range limits of proboscis monkeys in Sabah. This is not unlikely as the proboscis monkey range from the coast (according to research in Kalimantan) can be greater than 300 kilometres and as far as 750 kilometres, along the courses of major rivers. It is, however, not expected that large populations persist in those areas, and may be due to historical forest fragmentation, leaving small inland populations isolated from major populations nearer the coast.

The extant Sabah population is highly fragmented, with only five major centres of continuous distribution and numerous small isolated populations. The Klias Peninsula population is the only major centre of proboscis monkey distribution on the west coast and is separated from the east coast populations by the highland areas of the Crocker Range. On the east coast, populations in Tangkarason, Paitan, Sugut and Beluran appeared connected by coastal mangroves from the west of Pitas up to the Samawang area in Sandakan. Populations in Sandakan Bay and Kinabatangan are likely continuous along narrow coastal mangrove strips. The population along the Segama River is probably completely isolated. A small population is, however, still found in Kulamba Forest Reserve that links Segama to Kinabatangan. Tawau Bay has a continuous population along the extensive mangrove habitat, which is possibly connected with the major population of the delta of the Sesayap, Sembakung and Sebuku rivers in Kalimantan.

Female proboscis monkey with
infant more than a year old.

06

ACTIVITY AND BEHAVIOUR

One of the display behaviours in proboscis monkeys: adult male threatening unfamiliar object (other proboscis monkey group, predator, human, or other animals).

Proboscis monkeys are diurnal primates whose activities are closely associated with waterways. They sleep along water edges and travel inland to forage during the day, returning to their sleeping sites in the evening. Proboscis monkeys spend most of their time sleeping in the trees at night, normally starting to roost by 1730 hours and waking up at 0630 hours. Diurnal activity rhythm also indicates the majority of time is spent resting (77%), other than feeding (20%), and activity seems to be influenced by various temporal factors, for example, food availability, weather and air temperature. Resting behaviour tends to decrease from 1500 hours to 1700 hours and their feeding behaviour increases during this time, shortly before sleeping.

Adult male resting in the shade high up on a tree in the afternoon.

Depending on weather conditions, such as a rainstorm, the monkeys reduce their daily travel path or stay in the vicinity of their sleeping sites and start to roost much earlier at 1530 hours and move out from their sleeping sites after 0800 hours. Groups near coastal areas, where it tends to get bright earlier and dark later, tend to move into the forest earlier in the morning and move back to their sleeping sites later in the evening and this movement may also follow receding tides.

Proboscis monkeys are social primates displaying a variety of typical intra-specific behaviours, including agonistic vocalisations and stances, affiliative grooming, allo-mothering, play, sexual behaviours and vigilance, as well as a wide range of boisterous vocalisations —

Young male in threatening display.

Grooming between adult females in captivity, Lok Kawi Wildlife Park, Kota Kinabalu, Sabah.

Juvenile proboscis monkeys play-wrestling on the ground in a mangrove forest.

Mother and infant taking a rest.

Self grooming.

five types of 'growls', two types of 'honk', and seven types of 'shrieks' — indicating their social complexity. Interactions among individuals are observed both along river and inland forests, but interaction among groups is mostly at their sleeping sites along rivers in the late afternoon and early morning. They spend only about 0.5% of the daytime on social behaviours. Of the social behaviours, grooming was mostly observed between adult females and between adult females and their juveniles or infants. Female–male and male–male grooming is very rare in the wild but has been observed in captivity.

07

SOCIAL ORGANISATION

Proboscis monkeys are typically organised into stable one-male groups consisting of an adult male, several females and their offspring, as well as all-male groups consisting only of males. Some researchers have also indicated a multi-level society, a secondary level of association with fission-fusion of stable one-male groups within bands. Travelling along the rivers of Sabah, one can often see groups of proboscis monkeys in close proximity. Although both one-male and all-male groups have overlapping sleeping sites, these groups do not appear to form a high-density banding like some non-human primate species that have been shown to exhibit complex multi-level societies, for example, the geladas (*Theropithecus gelada*), hamadryas baboons (*Papio hamadryas*) and snub-nosed monkeys (*Rhinopithecus avunculus, R. bieti, R. brelichi,* and *R. roxellana*). There is no significant evidence of spatial clumping of proboscis monkey sleeping sites on a finer scale.

The aggregation patterns of proboscis monkey groups along rivers may be explained by the spatial heterogeneity along the river and influenced by the temporal change of food availability. Groups aggregate at places where the temporal food sources like fruits or flowers are more abundant although these folivorous monkeys are primarily dependent on leaves, which are abundant throughout the year. In addition, aggregation patterns may be influenced by the forest types

An adult male proboscis monkey leads a one-male group.

Proboscis monkey group size and density

Proboscis monkey group sizes in Sabah average 12.4 individuals (range = 1–28). One-male groups average 18.7 individuals (range = 2–28) and all-male groups average 9.2 individuals (range = 5–19). Mean group sizes of proboscis monkeys differ in different habitats. The size of one-male group is found to be higher in riverine forest (mean = 13.2) than freshwater swamp forest (mean = 12.3) and lowest in mangrove forest (mean = 11.5). Mean group size is related to densities in different habitats. This indicates that habitat type is an important determinant of group size and density although site-specific spatial distribution of resources, habitat quality and disturbance may also contribute to overall group size in a given area.

The basic social unit of the proboscis monkeys is one-male groups which typically comprise an adult male, several females and their offspring.

Infant proboscis monkeys are highly dependent on their mother for up to two years.

Aggression from dominant male to maturing males is often observed in captivity.

Female proboscis monkeys are known to disperse from their natal group.

that proboscis monkeys inhabit. In Sabah, proboscis monkey densities are higher in riverine forest than in mangrove forest, and inter-group distances at sleeping sites are shorter for riverine forest than mangrove forests. This may be due to higher food availability and spatial clumping of food resources in riverine habitats that result in higher aggregation of proboscis monkey groups in these habitats compared to those in mangrove habitats. Proboscis monkeys also aggregate along rivers where the river width is narrower because narrower river widths may provide good arboreal escape routes when landwards predators like clouded leopards attack the proboscis monkeys. It is also advantageous to the monkeys for predation avoidance of water-based predators like crocodile attacks when they cross the river. From tall trees at the edge of the water, the monkeys need only to be vigilant of predators coming from inland, and not from the riverside. The water below the sleeping trees provides easy escape for the monkeys when in danger. Apart from these reasons, there is anecdotal evidence that proboscis monkeys gather along riverbanks to keep themselves cool as they expend high energy levels for digestion, due to their complex sacculated

Extra group males disperse to form all-male groups.

forestomachs and large body size. Nonetheless, our investigations reveal that air temperature only had a small effect on the proboscis monkeys' preference for the riverine habitat.

Dispersal

Proboscis monkeys are female-centred, with the relationship between females (and their offspring) holding social groups stable though the frequency of social interactions among females is much lower than those of female-philopatric/male-dispersal primate species, e.g., macaques. Male and female proboscis monkeys are known to disperse from their natal one-male groups to all-male groups and other one-male groups, respectively. Some studies of dispersal in mammals and non-human primates indicate food resource constraints, or infanticide as possible factors for dispersal, but studies on proboscis monkey female dispersal suggested that these are unlikely factors in the case of proboscis monkeys. Female dispersal in proboscis monkeys occurs both in maturing females and mature resident females and, in addition, mostly occurs in non-aggressive contexts, unlike in some other primates, where vigorous attacks by males often occur. Aggression from the dominant male as well as mature adult females towards maturing males has, however, been observed in captivity.

Territoriality

Proboscis monkey groups seldom interact at sleeping sites, and physical intergroup hostility is absent or inconspicuous, although dominant displays by males are frequently observed. During a 13-month study of a proboscis monkey group at Kinabatangan, inter-group interactions were only observed three times in inland forest and in all the encounters an adult male in the study group drove off the rival alpha male, suggesting that the aggregation patterns among proboscis monkey groups along the river are not maintained in inland forest. Evidence indicates that proboscis monkeys are not territorial, and that it is not reasonable for a group to be able to defend its home range, although they have a preferred core area in their home range.

Interaction with Other Primates

Sympatric primate species often refuge together as a predator avoidance strategy. Proboscis monkeys are highly tolerant of other sympatric diurnal primates like the long-tailed macaque, the pig-tailed macaque, the maroon langur and silver langur that share their habitats and often sleeping sites or even the same trees. Proboscis monkeys are, however, at times displaced at sleeping sites by the langurs and macaques. Food competition may influence the interactions between proboscis monkeys and other sympatric primates, particularly langurs where food competition may be high due to similar digestion systems and thus preferred food items. However, subtle dietary divergence or high food availability in certain areas or during certain seasons may prevent antagonistic behaviour between proboscis monkeys and other primates.

Maroon Langur (*Presbytis rubicunda*).   Silver Langur (*Trachypithecus cristatus*).

Pig-tailed macaque (*Macaca nemestrina*).   Long-tailed macaque (*Macaca fascicularis*).

There is no evidence suggesting
territoriality in proboscis monkey groups.

08

FEEDING ECOLOGY

P roboscis monkeys are colobines that in nature feed exclusively on leaves, unripe fruits and seeds. The colobine stomach is characterised by enlarged and sacculated forestomachs for bacterial and enzymatic digestion of these hard-to-digest plant materials. Natural diets contain large amounts of leaves, and they have very long retention times because of the need to break down fibre using cellulose-digesting bacteria and deactivate toxins in leaves and seeds. These physiological adaptations allow colobines to utilise both young and mature leaves in greater quantities than most other sympatric primates. However, foods with high sugar content, such as ripe fruits, are readily fermented by bacteria and can result in rapid build-up of gases and acid in the stomach. Acidosis has been reported to be fatal for colobines. Nonetheless, the specific stomach structures of the different colobine species are slightly different — some of them have a marked preference for fruits (mostly unripe) or seeds rather than mature and young leaves.

The stomach of the proboscis monkey has the smallest area relative to total gut and the small intestine is relatively the largest of the colobine species, indicating their low fermenting capacity and high absorbing ability, which aids its more frugivorous diet.

The main diet of proboscis monkeys are leaves.

Food Habits and Dietary Diversity

In Kinabatangan, Sukau, proboscis monkeys have been observed consuming a total of 188 plant species (127 genera, 55 families). This is more than the 90 species in 39 families reported from earlier research. Young leaves (65.9%) and fruits (25.9%) accounted for the majority of feeding time in Sabah; this is comparable to long-term studies in Samunsam and Tanjung Puting National Park, indicating young leaves, fruits, and flowers constitute 38–52%, 40–50%, and 3%, respectively, of the diet. These results show that the feeding ecology of proboscis monkeys is more diverse than previously thought.

Proboscis monkey feeding on leaves.

Fruit Eating

Colobine monkeys are unusual among primates in exhibiting a preference for unripe fruits over ripe fruits. Over 90% of fruit feeding by proboscis monkeys involves the consumption of unripe fruits. They spend 32.9%, 2.1% and 0.9% of their fruit-eating time on consuming seeds, flesh and calyxes, respectively. When the seeds and the flesh of the fruit are firmly stuck to each other, proboscis monkeys consume both the seeds and flesh concurrently in most cases (64.1%). Consumption of seeds makes up 97% of their fruit-eating time and when the monkeys collect ripe fruits they almost always abandon the flesh and only consume the seeds, suggesting that fruit foraging is targeted at obtaining seeds for consumption.

Eugenia litseaefolia.

Dillenia excelsa.

Mallotus muticus (most dominant food tree eaten by proboscis monkeys).

Lophopyxis maingayi (most dominant food vine eaten by proboscis monkeys).

Tree Bark and Termites

Tree bark, termites and termite nest materials are consumed by proboscis monkeys. They consume the bark of papery bark tree species like the *Eugenia* sp., but not *Pternandra galeata* although this species is more abundant in our study site in the Kinabatangan area. Although many termite species are present within the proboscis monkey habitat, they only consume termites and nest materials of *Microcerotermes distans*, an arboreal species. As Asian colobines rarely consume animal matter due to their specific stomach structure, it seems unlikely that proboscis monkeys consume termites as a protein source. Consumption of bark and termites is likely to supplement mineral intake, as a buffer to forestomach pH or to aid in absorption of toxins. Maroon langurs (*Presbytis rubicunda*) in Sabah are known to feed on termite nests (*Macrotermes* sp.). Howler monkeys (*Alouatta belzebul discolor*), which is a leaf eater similar to the proboscis monkey, are also known to consume tree bark or soil when consumption of leaves increases considerably. Soil eating or geophagic behaviour has also been observed in captive proboscis monkeys.

Tree bark (*Eugenia* sp.).

Geophagic behaviour in proboscis monkeys.

Termite nest.

Apart from feeding, proboscis monkey spend most of the daytime resting in the trees.

Drinking

Colobine monkeys, including proboscis monkeys, obtain water primarily from leaves during digestion of cellulose in their stomach and rarely drink water. Nonetheless, proboscis monkeys have occasionally been observed drinking water; they use the rivers, small streams, or temporary pools as water sources. This is most common for adult males, while juveniles almost always restrict themselves to hand-dipping water from tree holes without descending to the forest floor to utilise other water sources. Adult females have been observed to use both methods. These differences in drinking behaviours among the age-classes suggest that preferred methods of obtaining water may be related to predation avoidance strategies. As smaller sized monkeys are more vulnerable to predators, they prefer to obtain water from safer places (in trees but not on the ground).

Relation between resting and digestion

In order to obtain sufficient nutrients, proboscis monkeys feed throughout the day. Apart from feeding, they spend most of the daytime resting. Proboscis monkeys devote a large proportion of their activity budget to resting (77%), similar to other colobine species (e.g., *Trachypithecus francoisi*, 52%; *Presbytis thomasi*, 59%; *Colobus vellerosus*, 59%; *C. guereza*, 63%) to aid in the digestion of foods with high levels of cellulose. Feeding trials of captive proboscis monkeys support this suggestion because food transit and retention times are long (14 hours and 52 hours respectively). Furthermore, studies have shown a positive relationship between time resting and body size among African colobines. Proboscis monkeys, which are the largest Asian colobines, appear to fit this trend.

Variation in Diet

Proboscis monkey feeding behaviour may vary with temporal fluctuations in environmental factors. Daily movements out of riverine forest and dipterocarp/high kerangas forest into mangrove trees (where intensive feeding occurred) in the afternoon, followed by movement back into high forest types (for resting) in the evenings have been observed in Samunsam. Seasonal shifts between habitats used for feeding have also been observed, possibly indicating that some habitats may be preferred; during some times of the year when preferred food resources are scarce, they may utilise other habitats for feeding. In Sabah, proboscis monkeys also showed marked seasonal fluctuation in time spent feeding. Flower-eating is constantly low throughout the year, while young leaf-eating was almost constantly highest. Fruit-eating, however, was low in some seasons, but in the fruiting seasons was significantly higher than the percentage of time devoted to young leaf-eating. The number of plant species consumed per month by proboscis monkeys also fluctuates seasonally, ranging from 36 to 82. Proboscis monkey dietary diversity increases when preferred foods (fruits) are scarce. The monkeys increase daily travel distances during months in which they predominantly feed on young leaves. Increased travel and dietary diversity when consuming young leaves may serve to reduce the potential risk of toxin accumulation associated with feeding on a limited number of plant species.

A pet monkey fed on a diet of vegetables and milk for six years appeared in good condition.

Proboscis monkey feeding on the inflorescence of a coconut tree.

However, during unripe fruit-eating periods, proboscis monkeys exploit a limited number of plant species. This may indicate the fact that unripe fruits eaten by proboscis monkeys usually contain fewer toxins than young leaves. Phytochemical analyses are needed to determine this hypothesis in future studies.

Dietary Flexibility

Contrary to earlier assumptions that colobines mostly exploit ubiquitous food sources such as leaves, several recent colobine studies have reported high levels of fruit and/or seed consumption in response to local conditions (for example *P. rubicunda*, *T. pileatus*, *T. auratus*, *C. angolensis*, *C. guereza* and *S. vetulus*). Our studies on proboscis monkey feeding ecology also provide evidence of dietary flexibility. Although the majority of feeding time is devoted to the exploitation of young leaves, during certain times of the year, unripe fruits and seeds accounted for more than 50% of monthly feeding time. In captivity

Sliced cucumbers and pancakes cut into bite-size pieces.

in Singapore Zoo, proboscis monkeys thrive on a diet of mixed leaf browse, vegetables, fruits, primate pellets and rice balls. In Sabah, a pet monkey kept for six years on a diet of vegetables, milk and the occasional prawn cracker appeared in good health. The diet of proboscis monkeys in Labuk Bay, Sandakan is supplemented with pancakes, cucumbers and long beans on a regular basis in addition to natural food items available from the surrounding mangrove forests. Records of proboscis monkeys persisting in small habitat fragments, rubber plantations and feeding on crop plants like coconut also support the theory that proboscis monkeys may have a higher amount of dietary flexibility than previously thought. In Pulau Kaget, the depleted forests could support a high population density because of adaptation to water plants as food resources containing high mineral content e.g. *Limnocharis flava* (Limnocharitaceae), *Agapanthus africanus* (Alliaceae), *Hymenachne amplexicaulis* (Poaceae) and *Vittis trifolia* (Vitaceae). However, the diet of proboscis monkeys is much less diverse than that of the more omnivorous orangutans and macaques, which actively forage on a larger variety of foods, including oil palm fruits and shoots, as well as human food sources.

09

RANGING BEHAVIOUR

Various environmental factors influence the ranging patterns of primates, but distribution/abundance of food resources and predation threats are the most important factors. If food resources are readily available, individuals do not have to range extensively in search of food. In general, the ranging behaviour of primate species that prefer to feed on limited and seasonal food sources, such as fruits and flowers, is more influenced by food distribution and abundance than that of primate species which feed on equally distributed foods, such as leaves. The daily travel distance is longer, and the home range is larger, when the diet is based on seasonal and limited food sources.

Daily Travel Distance and Home Range

Their daily distance travelled average 706–910 metres and ranged from 370 metres to 2 kilometres. The mean perpendicular distance from the river bank is between 400–750 metres. In Sabah, the daily travel distance of proboscis monkeys is 220–1734 metres, with a mean distance of 799 metres. In their daily travelling, the mean perpendicular distance from riverbanks is approximately 200 metres (range = 0–800 metres).

Based on studies at sleeping sites using boat surveys, proboscis monkeys are known to have home ranges varying from 0.3 square kilometre to 9 square kilometres and overlap by up to 95%. There is evidence that proboscis monkeys are relatively sedentary animals, with high habitat fidelity and limited ranging in the same areas, as observed in Bako National Park, Sarawak and Garama, Sabah over periods of up to four months. However, long-term movements of proboscis monkeys indicate movement of about 4.5 kilometres to more than 7 kilometres along rivers. In Sabah, observations of a one-male group of proboscis monkeys in inland riverine forest for 13 months indicated a home range of 1.38 square kilometres and the group used both riverbanks situated at a distance of 1400–5250 metres from the river mouth. Although the group travelled long distances when it moved along the Menanggul River, the daily travel distance was rarely more than 1000 metres when it travelled away from the river, evidently because the monkeys had to return to the riverside before sunset. The small range of movement is also reflected in their activity budget of an average of only 3.5% of their time spent locomoting.

Home Range Differences between Habitats

Home range size of non-human primates generally decreases when the food sources become more abundant. The home range size and per capita home range size of proboscis monkeys can similarly be explained by food availability, and differ between habitats. In mangrove forests of Abai, the home range was 3.15 square kilometres and per capita home range 0.22 square kilometres per individual. In mixed mangrove and lowland forests of Samunsam it was 9 square kilometres (0.56 square kilometres per individual), in peat swamp forest of Tanjung Puting — 1.37 square kilometres (0.13 square kilometres per individual) — and in riverine forest of Sukau — 2.21 square kilometres and 0.11 square kilometres

Home and daily range are related to food availability and distribution in different forest types.

per individual, and 1.38 square kilometres (0.09 square kilometres per individual). In mangrove and mangrove-based forests, extremely high home range values and per capita home range compared to riverine and peat swamp forests suggest that mangrove forests have low food availability compared to other forest types.

The daily travel distance of proboscis monkeys significantly correlates with the availability of fruits, but not with that of flowers or young leaves. The negative relationship between the daily travel distance and fruit availability indicates that the monkeys utilise a small home range in fruit-abundant seasons. This is unique amongst colobine monkeys as other Asian colobines, such as banded langurs (*Presbytis melalophos*), hanuman langurs (*Semnopithecus entellus*) and capped langurs (*Trachypithecus pileatus*), increase their travel distance in months when the diet is based on fruits or flowers. Proboscis monkeys feed on fruits of dominant plant species which are more easily available, and this may explain their smaller home ranges and shorter daily travel distance in fruiting seasons.

Jumping from tree to tree is part of the proboscis monkey's locomotory repertoire.

The clouded leopard (*Neofelis diardi*), a predator of the proboscis monkey.

10

NATURAL PREDATION

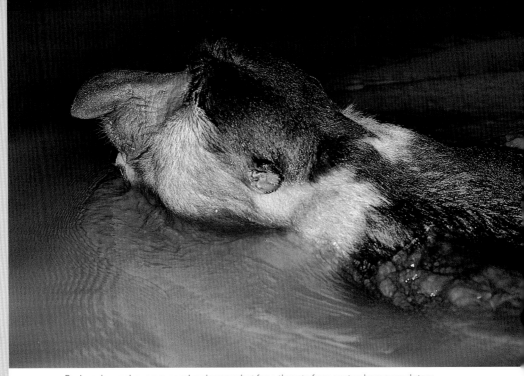

Proboscis monkeys are good swimmers but face threats from water-borne predators.

Natural predation has been reported to play an important role in the morphological and behavioural evolution of non-human primates. Adult males of proboscis monkeys have the largest body size among colobine species (*ca.* 20 kg), and predation is expected to be low because of a negative correlation between body size and predation pressure, and also because predation pressure on arboreal primates is reported to be lower than that on terrestrial primates. Nonetheless, proboscis monkeys are at risk of predation, and predation pressure may influence anti-predator and risk-sensitive behaviour in proboscis monkeys.

Reticulated python (*Python reticulatus*) preying on a proboscis monkey.

At least four long-term studies (12–30 months) on proboscis monkeys have been conducted in Borneo with only one case of predation reported — that of an adolescent female proboscis monkey attacked by a false gavial (*Tomistoma schlegeli*). Apart from these studies, one observation of predation by false gavial, four by clouded leopards and one case by a reticulated python have been reported. Estuarine crocodiles (*Crocodylus porosus*) are potential predators of proboscis monkeys, as they have been observed to prey on adult bearded pigs (*Sus barbatus*). Other potential prey species include reptiles such as blood python (*Python curtus*) and monitor lizard (*Varanus* sp.). Raptors, such as black eagle (*Ictinaetus malayensis*), crested serpent-eagle (*Spilornis cheela*) and bat hawk (*Macheiramphus alcinus*) may also prey on infants and juveniles and have been reported in long-tailed macaques. The predation events by clouded leopards (*Neofelis diardi*) observed in Sabah occurred during the day in trees. Thus, the ability of clouded leopards to travel and hunt in trees during both day and night is a potential threat to these primates. Although current available data are insufficient to evaluate the extent of the predation threat on proboscis monkeys, those in the wild are probably exposed to a high predation threat.

Estuarine crocodile (*Crocodylus porosus*).

Proboscis monkeys do not select their sleeping trees at random. Safety from terrestrial predators, such as the clouded leopard, is one of the factors affecting their choice of sleeping trees.

Characteristics of proboscis monkey's sleeping trees

Due to predation pressure, night refuging behaviour and sleeping tree selection of proboscis monkeys are related to anti-predator strategies. In Menanggul, Sabah, proboscis monkeys prefer to select sleeping trees where branch-to-bank distances are narrower since these places provide good arboreal escape routes from the clouded leopards. However, they occasionally select trees where branch-to-bank distances are large (greater than 10 metres). These sleeping sites are located in small trees adjacent to emergent trees and whose canopy and branch structure offer limited access to leopards. In Klias Peninsula where predation pressure is low, especially from terrestrial predators like leopards, sleeping trees are not strictly at riverbanks (but usually between 15 to 30 metres from riverbanks with a maximum recorded distance of 46 metres from riverbanks). Also there was no tendency to sleep on trees where branch-to-bank distances are narrower, suggesting that the monkeys in Klias can select their sleeping sites more freely compared to the monkeys in Menanggul, with high predation pressure. In fact, proboscis monkeys were frequently observed to travel on the ground in Klias, supporting the impression that predation pressure is low.

Compared to available trees in the forests in Klias, characteristics of sleeping trees of proboscis monkeys in Klias generally have large trunks (mean =143.6 cm girth at breast height), are tall (mean = 34.3 m), with many (mean = 6) large (mean = 24.1 cm circumference) main branches. The main branches are also angled at a large degree from the tree trunk (greater than 45°). Such trees are located near to other trees, with overlapping branches, creating good arboreal connectivity. Selection of sleeping trees is not dependent on the crown type and crown density. In fact, proboscis monkeys are often seen resting on almost bare trees. Large trees are probably important to support the weight of individual proboscis monkeys (adult males can reach up to 24 kilograms in weight), and several individuals often occupy the same tree. Since sleep occurs exclusively in a sitting position, the preferred features of the tree branches are likely comfort related as it is more comfortable for a monkey to sit on a large horizontal branch then a small and slanting one. The connectivity of a tree with other trees is important as the arboreal proboscis monkey travels from tree to tree using connecting branches.

Effects of Predation Threats on
River Crossings and Sleeping Sites

Due to predation threats, proboscis monkeys employ anti-predator strategies in river crossings to minimise the frequency and time spent crossing rivers. Observations in Sabah and Natai Lengkus, Tanjung Puting, showed proboscis monkeys more frequently select locations with narrower branch-to-bank distances, where the river is narrower and where the arboreal canopy extends across much of the river. Adult males use arboreal routes to cross the river more commonly than adult females or juveniles. This may reflect the fact that larger-bodied adult males can leap longer distances and go directly from tree branch to branch without entering the water. Smaller-bodied females (often with an infant) and juveniles may have more difficulty spanning this distance and therefore are forced to swim across rivers more frequently. In Bako, Sarawak, proboscis monkeys refrain from swimming at places where crocodile populations are high.

As an anti-predator strategy against crocodiles, proboscis monkeys have been reported to cross rivers at locations where the river is narrow. The monkeys may cross the river either by leaping from the branch of a tree on one side of the river bank to the branch of a tree on the other river bank or by swimming (see inset).

11

THREATS TO
PROBOSCIS MONKEYS

Proboscis monkeys are commonly seen along rivers due to their habit of sleeping along waterways, and this gives a general impression that the species is common and unthreatened. Recent studies in Sabah, as well as evidence from investigations in other range countries, however, show that the proboscis monkey is now highly threatened.

Habitat Loss

Borneo, the only home to proboscis monkeys, is estimated to have only 54% of natural forest cover remaining and, judging from the trend of forest loss, only 23–32.6% of Borneo will be covered in natural forest by 2020. Sabah's economy is still basically agrarian and natural resource-based and is dependent on the production and export of natural resource commodities, a contributor to natural forest cover being reduced from an estimated 86% in 1953 to 57.4% by 2001. The effect of habitat loss is amplified in the case of proboscis monkeys due to their habitat specificity to riverine, swamp and mangrove habitats, which also happens to be the habitats most used by humans. In Kalimantan and Sarawak, decline of proboscis monkey populations can mainly be attributed to habitat destruction caused by human activities, especially in riverine forest and coastal zones. Similarly, human land use pressure through expansion of human habitation, agriculture and other land developments is diminishing the most important riverine forest and

Oil palm plantation by the river bank replacing riverine forests.

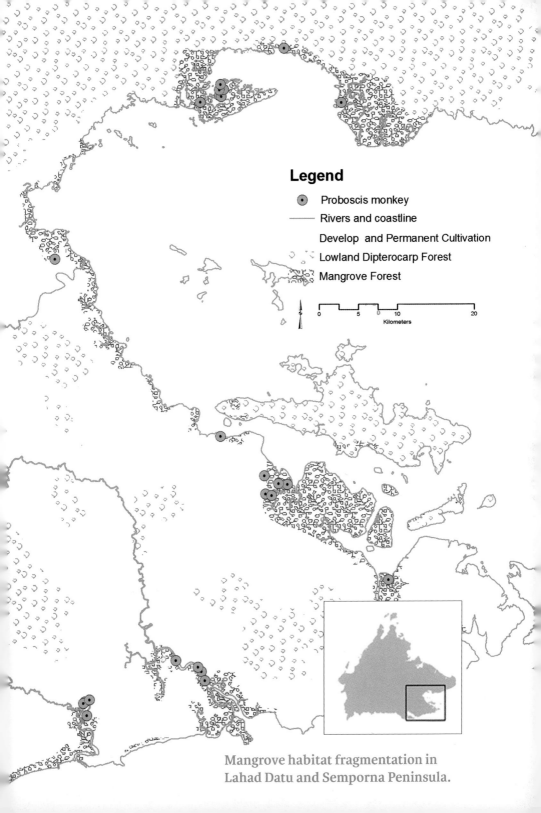

Legend

- ⊙ Proboscis monkey
- —— Rivers and coastline
- Develop and Permanent Cultivation
- Lowland Dipterocarp Forest
- Mangrove Forest

0   5   10   20
Kilometers

Mangrove habitat fragmentation in
Lahad Datu and Semporna Peninsula.

coastal mangrove habitats for proboscis monkeys in Sabah with only an estimated 9.8% of suitable proboscis monkey habitats remaining. The largest proboscis population in Kinabatangan — about 25% of the total population in Sabah — is surviving in only an estimated 0.7% of the total forested area.

Apart from direct human causes, forest fires and droughts brought about by climatic changes, for example, the El Ninŏ event of 1997–98, have caused proboscis monkeys to lose a greater percentage of their remaining habitat than any other primate species in Borneo. Sabah experienced six periods of drought in the last two decades with the forest fires of 1983, 1986 and 1997–98 in particular causing serious forest loss in the Klias and Sugut areas where major proboscis populations live.

Habitat Fragmentation and Degradation

As a consequence of localised habitat loss, proboscis monkey habitats are increasingly fragmented and this is becoming an emerging concern for proboscis monkeys. In Sabah, habitat fragmentation is most evident along major rivers like the Segama, Sugut and Kinabatangan, and coastal areas in Sandakan, Lahad Datu and Semporna. Proboscis monkeys are predominantly dependent on riparian and coastal habitats for food and shelter and for an arboreal, habitat specialist, intervening matrixes of cultivated land, human settlements or grassland areas between their preferred habitats, especially along riverbanks, can have many undesirable immediate and longer-term consequences. Proboscis monkeys do not use areas of extensively burnt dryland forest areas or permanent agriculture like oil palm or crop land, unlike

The forest is cleared and the soil terraced to make way for oil palm cultivation.

more omnivorous species such as the macaques and orangutans. Food resources in forest fragments are limited and this results in resource competition with other primate species. Extensive use of waterways by humans and forest conversion and extraction through logging results in pollution and further degradation of already small and fragmented proboscis monkey habitats. Social activities are also impeded if groups do not have the opportunity for interactions with other proboscis monkey groups. Effective population size is reduced and gene flow disrupted, resulting in greater likelihood of local extinctions. Such longer-term effects of fragmentation may not be evident now as the initial response to habitat shrinkage is for survivors to crowd into remaining forest habitats, causing an increase in population density before the effects of over-crowding and habitat saturation take place. Unusually high densities of proboscis monkeys are already evident in many areas. In Garama, Sabah and various other localities, large numbers of proboscis monkeys are found in narrow strips of forest that are, in places, less than 20 metres in width. Mahakam Delta, Kalimantan, which had extensive mangroves and tidal swamps up until the early 1990s, now has only a few remaining fragments, which are overpopulated with proboscis monkeys. Increasing loss and fragmentation of once continuous areas, coupled with other direct and indirect human and natural impacts, are critically threatening the continued existence of proboscis monkeys which depend on these forests for their survival.

Habitat loss and fragmentations has left tiny strips of important forest habitats for proboscis monkeys.

Hunting

The proboscis monkey's generally large body mass and habit of sleeping along rivers make it an attractive and easy prey for hunters. While hunting primates as a food source is common in Africa, proboscis monkeys are not exactly highly valued for their meat in Asian cultures. In Kalimantan, bezoar stones in the intestines of proboscis monkeys are valued and used in traditional Chinese medicine and the penis of the male proboscis is used as an aphrodisiac. In Kalimantan, it was noted that the advent of guns and outboard motors coincided with a decrease in proboscis monkey abundance as it increased access to hunting in

Due to their close association with riverbank habitats, large numbers of proboscis monkeys can easily be tracked down by hunters in boats.

coastal areas. Historically, hunting in inland areas in Kalimantan, especially in areas inhabited by non-Muslim Dayaks, has reduced populations in inland areas or exterminated populations in some areas of otherwise suitable habitat. In Sabah, the main distribution of proboscis monkeys coincides with areas with dominantly Muslim groups who traditionally do not hunt monkeys. Although locals are usually not involved in direct hunting activities, they may facilitate hunting by renting boats and imparting knowledge about the location of proboscis monkeys.

Habitat loss and disturbance as a result
of the timber industry is one of the threats
faced by the proboscis monkey.

12

CONSERVATION OF PROBOSCIS MONKEYS

Proboscis monkeys are legally protected across their entire range and classified as "Endangered" by the World Conservation Union (IUCN). The species is also listed under Appendix I of the Convention of International Trade in Endangered Species of Flora and Fauna (CITES) which prohibits all commercial trade of the species. In Sabah, they are also fully protected by the Wildlife Conservation Enactment (1997). Judging from the threats that the proboscis monkeys are facing today, urgent mitigating actions are required to safeguard the long-term survival of the species.

Protected Areas and Networks

Current distribution of proboscis monkeys in Sabah appears disjunct with many highly fragmented populations and large populations are themselves fragmented into sub-populations by various forms of land conversion. While long-term effects of habitat loss and fragmentation on proboscis monkeys are still largely unknown, protected areas are important for ensuring the integrity of sufficiently large tracts of continuous forests with sizable healthy and genetically diverse proboscis monkey populations.

(Left & right). A proboscis monkey, illegally kept as a pet.

Proboscis monkeys in semi-wild conditions.

The current forest reserves network in Sabah harbours only 15.3% of proboscis monkeys in Sabah. It is important to extend protection status to currently unprotected areas with important proboscis monkey populations and confer totally protected status in the form of national parks or wildlife sanctuaries to areas with major populations. The habitat specificity of proboscis monkeys requires special consideration when determining effective protected areas providing suitable habitat links between important populations. For example, an extension conservation area linking three small reserves containing proboscis populations within the Klias Peninsula was successfully gazetted as the Bukau-Api Api Protection Forest Reserve under a United Nations Development Programme/Global Environment Facility (UNDP/GEF) funded peatswamp forest project. More efforts are needed to preferably link major populations through a protected area network, as part of a conservation strategy that extends beyond borders for the protection of the species across its range.

Forest corridors are important for proboscis monkeys to allow uninterrupted movement from one part of the forest to another along the riverbanks.

Re-establishment of Forest and River Corridors

As a result of overexploitation of natural forests, extensive rehabilitation activities are now necessary, as reflected in the Sabah Forestry Department's strategic plan for forest resource development. Riverine habitats in particular are not sufficiently protected, compounded by the fact that most human activities occur along rivers. Along the Kinabatangan and Segama rivers, plantation and logging activities have reached the riverbanks. Although the Sabah Water Resource Enactment 1998 stipulates the preservation of 20 metres of riverine forest perpendicular from rivers, this is insufficient as proboscis monkeys range daily up to 800 metres perpendicular distance from the riverbank inland to forage. To ensure the integrity and effectiveness of protected areas for proboscis monkeys, the retention of wider linear remnants along rivers is necessary, as well as the restoration of degraded habitats. It is encouraging that there is impetus by the State Government of Sabah to identify the severity of the problem of riparian reserve encroachment along state rivers by planters, and plans to rehabilitate riverine forest along rivers that have been converted to oil palm plantations. Greater efforts must be dedicated to restoring remnant habitat patches degraded by agriculture and other human activities, as well as re-establishing corridors along fragmented river systems.

Translocation of Highly Isolated Populations

Increasing loss of suitable proboscis monkey habitats is resulting in many remnant populations facing the threat of local extinction. The local extinction of proboscis monkey populations in Sabah as a result of habitat loss has been recorded in Papar and in Kunak. This is likely to be only the tip of the iceberg, with many other populations disappearing unrecorded. Elsewhere, local extinction of populations of proboscis monkey as a result of habitat loss has been reported on Pulau Kaget in Indonesia. The population in the now famous Labuk Bay Proboscis Monkey Sanctuary which was almost exterminated by forest clearance for oil palm plantations indicates that an interventionist approach may be needed to protect certain at-risk populations. For highly isolated proboscis monkey populations which are likely to go locally extinct without intervention, translocation seems the only logical step. Translocation of sensitive primate species populations remains highly controversial due to associated high mortality rates, introduction of diseases or parasites and disruption of food resources and social structure equilibriums. An example can be found in the difficulties that were encountered during the translocation scheme of the Pulau Kaget Nature Reserve population. Preferred reintroductions into areas where there are no remnant populations are not entirely feasible due to the rapid loss of suitable proboscis monkey habitats. Monitoring of isolated populations in unprotected areas, with consideration of translocation into areas with existing populations, coupled with proper post-release monitoring, or to captive establishments may be the only feasible options.

Community Involvement in Conservation Activities and Awareness

As the economy of local human populations is invariably linked with resource extraction from habitats of the proboscis monkey, provision of alternative incomes and capacity building for more wildlife-related

Involvement of the local community in ecotourism, conservation and research activities is an important part of long-term conservation programmes for the proboscis monkey.

work like enforcement, ecotourism and research that benefits local communities and proboscis monkeys will aid long-term conservation of the species. Research, ecotourism and enforcement initiatives are important components for local involvement as they are usually knowledgeable about their areas and the animals, and this knowledge can be easily harnessed with proper training. If conservation initiatives for proboscis monkeys are to achieve long-term success, educational initiatives targeted at all levels from local villagers to landowners to resource-managers to policymakers are needed as the human factor is an important element in conservation planning. Public awareness by all strata of society will be an important component for effective implementation of conservation plans related to the proboscis monkey, and information that is collected should be made readily available to the local communities and industries in the form of conservation awareness programmes and the promotion of conservation-friendly ethics and practices. A multi-species, community-based conservation approach providing both academic and practical management aspects will benefit conservation of all inhabitants and habitats in areas otherwise threatened by habitat loss and encroachment.

With the inevitable trend of wildlife habitat loss expected to continue, zoological institutions will play an increasingly prominent role in the conservation of endangered species threatened by extinction. Zoos have in the past had limited success in maintaining captive populations of sensitive leaf-eaters like the proboscis monkey. They are notoriously difficult to keep in captivity due to their special diet and habitat preferences. The first pair of proboscis monkeys was sent to the San Diego Zoo in 1956 and there was a subsequent influx of proboscis monkeys into zoos in Europe and America. However, none of these zoos was able to maintain viable populations and by 1997, only Bronx Zoo had proboscis monkeys left, with the last two eventually sent to Singapore Zoo in 2003. In its range countries, Surabaya Zoo, Yogyakarta Zoo, Ragunan Zoo and Taman Safari in Indonesia and Lok Kawi Zoological and Botanical Park in Sabah are known to be holding proboscis monkeys. Surabaya Zoo has one of the largest collections (~55 individuals). Outside its range country, the Singapore Zoo has the largest collection (20 individuals) with 19 successful offspring from its captive population. Yokohama Zoorasia received five monkeys from Surabaya Zoo in June 2009 and is the only zoo outside Southeast Asia to be currently holding proboscis monkeys. With proper management practices, it is feasible to maintain captive stock in zoological institutions.

Understanding the behaviours of animals in the wild is an essential step towards better management of captive populations and captive information can in turn aid in research and management of wild populations. Captive-based research has contributed to important knowledge on proboscis monkeys that could not be obtained from wildlife research, for example, aspects of diet, nutritional, disease and life history information. Establishing a well-strategised *in-situ* and *ex-situ* collaborative link would also provide the necessary impetus for conservation education and awareness-related initiatives which will be beneficial to the long-term conservation of the species.

Some of the problems related to inadequate species conservation are due to the lack of research data and longer-term monitoring efforts that can be applied to sound conservation and management decisions. The link between research, policy and implementation needs to be better established so that important information can get to the level of important decision makers and other stakeholders and not simply be for academic interest. Much research on proboscis monkeys to date has been short-term efforts concentrated in a few localities of easy accessibility and largely uncoordinated in terms of research methodologies and priorities. The setting up of permanent community-based field stations at important sites like Sugut and Segama where a few studies have been conducted to date will enable longer-term studies which are important to collect longitudinal data.

The following research priorities are recommended:

1. Long-term continuous monitoring of basic data on population numbers and distribution for important and threatened populations and standardisation and improvement of rapid census methods for related studies across proboscis monkey range;
2. More accurate and updated habitat and land use classifications important for more detailed habitat, land use pattern analyses;
3. Habitat and site-specific demographics, home-range, activity budgets, floristics and feeding ecology to determine important factors that contribute to the long-term viability of populations.
4. Population genetics to elucidate genetic status and possible effects of fragmentation.

The Future is Now

The information presented in the preceding chapters of this book represents the bulk of what we know about the proboscis monkey to date. Novel findings included a more extensive range of proboscis monkey habitats than previously thought (they are not a pure mangrove species and use extensive riverine, freshwater and peat swamp forests); a larger diversity of food items (188 species, twice what was previously recorded and includes extensive seed and unripe fruit eating for what

Is there a future for the remarkable proboscis monkey?

was thought to be a predominant leaf eater) and dietary flexibility (including seasonal feeding variations); a unique social structure that is dependent on the spatial distribution of food resources; proof of a wider range of predators which includes the clouded leopard and reticulated python; and strong evidence of habitat loss and fragmentation along rivers that is impacting the proboscis monkeys that are strongly reliant on riverside forest habitats. Despite these findings and a wealth of previous decades of research, there still remain many knowledge gaps, as discussed earlier in this chapter; some of which are imperative to the long-term conservation of this enigmatic species.

On-going and future research will no doubt uncover many more important answers in the near future and contribute to the wealth of knowledge. Most of these answers that we will uncover in the future are not likely to be pleasant, judging by the trends of continued damage that humans inflict on wildlife and wild habitats that result in species endangerment and extinction in general. The threats to proboscis monkeys we see today, in the strictest sense, are not novel and can be seen as a mere extension of human impacts that have been accumulating over many years. Hunting for food and clearance of forests for cultivation and human habitation have ensued since the advent of modern man and some of these activities still remain a vital

The long-term survival of the proboscis monkey depends on all of us.

part of human societal advancement and sustenance. The key issue therefore is how far are we away from a tipping point, where the negative impacts of human actions on proboscis monkeys become irreversible and mitigating measures would prove a tad too late?

Even professional conservationists are regularly confronted by the issue of balancing pessimism, optimism, and realism in the course of their chosen work. At times, the more we learn or are in the know of the actual situations and depressing plight of species on ground zero, the more likely is the development of a sense of minuteness, hopelessness and fear that there may be only so little we can actually do. Conservationists can, and often do, seek solace in the perception or actuality that we are doing what we can to contribute to our adopted mission, but for the man and woman on the street, who may share our values, but may not be able to make direct contributions, the situation must seem even more exasperating.

Although specialised research and conservation strategies for the proboscis monkey are important in providing possible solutions for better conservation action, these efforts are futile without wider-ranging support. The clichéd but timeless phrase "every little effort counts" holds true and the future for proboscis monkeys is now! Simple lifestyle changes and choices towards more sustainable living or direct efforts to conserve species, however minuscule they may be, will contribute to a paradigm shift towards the development of a consciousness and awareness of the impacts human actions have on wildlife and cultivating a respect for the existence of other species besides ourselves, and the importance of this co-existence. Hope exists that enough will be done that will take effect in time to ensure the long-term survival of the proboscis monkey.

In the course of our research work, we have witnessed the importance of basic human traits needed for co-existence between humans and wildlife; and their contribution to conservation, as seen along Sabah's rivers of life. The fact that the major concentrations of proboscis monkeys in Sabah today are in areas where the local community do not traditionally hunt or eat monkeys, and the numerous successful and expanding community-based conservation initiatives for the proboscis monkeys today, provides important insights into the interconnectedness of humans and wildlife and how humans can positively impact the conservation of species. In the production of this book, we hope we have achieved an important objective — besides sharing knowledge about the natural history of proboscis monkeys — to encourage a wider appreciation for the species and evoke a stronger impetus for urgent conservation action to safeguard the heritage of this unique species beyond the conservation community and to the public domain. In the next and final chapter, we present information on places to visit in Sabah to view and appreciate proboscis monkeys, in a sustainable way that minimises disturbance to these sensitive animals.

The sequel to the proboscis monkey's story is eventually up to all of us!

Adult male proboscis monkey looking to its survival for the future.

13

ECOTOURISM

Proboscis monkeys in Labuk Bay Sanctuary in Sandakan.

The proboscis monkey is a highly charismatic primate species which has received increasing attention in recent years. In Sabah, proboscis monkeys have become an important flagship species for conservation, as well as an ecotourist attraction possibly of parallel importance to the orangutans. Areas where there are large and easily sighted populations, for example Garama and Sukau, have seen the proliferation of large- and small-scale tourist establishments over the last 5–10 years to cater to the increasing tourist volume. Although community-based ecotourism can bring significant benefits, for example, alternative income as incentives for local communities and policy makers to protect the species in areas of interest, the potential benefits gained from ecotourism are frequently counteracted by risks of exposure to unique pathogens, poor adjustment to novel environments, and changes in natural dispersal patterns. Proper evaluation and control mechanisms for ecotourism are needed to prevent irresponsible mass tourism activities by profit-oriented establishments that are detrimental to the long-term conservation of sensitive species. Any ecotourism initiative has to be sustainable, minimise disturbance effects on proboscis monkeys and integrated with sound conservation education programmes for the general public, capital stakeholders and the local community.

Sabah is certainly one of the best places to visit to get an opportunity to see proboscis monkeys in their natural habitats. Some places where

Proboscis monkeys are given supplementary food at the feeding platform daily.

proboscis monkeys are found are easily accessible and served by good communication and transportation systems. The tourism sector in Sabah is well established, so arrangements can easily be made with local tour operators based in major cities and towns like Kota Kinabalu. Most tour companies also maintain informative websites which potential visitors can access from anywhere in the world.

Unlike many wildlife species, such as the Bornean orangutan and Bornean elephant, the proboscis monkey is an easily observable animal with high encounter rates in the wild. In fact, at some places, seeing proboscis monkeys is guaranteed at any time of the year! This is due to their habit of sleeping along waterways. Observing proboscis monkeys in the wild is thus usually by boat. Although the best time to see proboscis monkeys varies slightly from place to place as access depends on local tide conditions and other factors, in general it is best to observe them in the early morning just after dawn (before the animals move into the forest from their sleeping sites), and in the two hours before dusk (when the animals move back to the riverbanks to start their preparations to settle down for the night). The latter is generally more popular amongst tourists as the timing is more convenient, as well as the extra attractions offered by the forest at these hours, for example, the spectacular views of sunsets and fireflies.

Guidelines for Observing Proboscis Monkeys

Increasing proximity of humans to proboscis monkeys as a result of eco-tourism may result in disturbance and altered behaviours that are detrimental to the long-term conservation of the sensitive proboscis monkeys. For example, the presence of a large number of tourists in boats on the river often prevents the monkeys from crossing rivers, and may therefore deny the animal access to vital food sources. Several rules and regulations are recommended for both tour operators and tourists when observing proboscis monkeys in the wild.

1. Tourists are advised not to make any loud noises deliberately, including noise emitted from electronic devices such as hand phones, audio players and portable video games.
2. Tourists and operators should refrain from making provocative gestures in any manner to attract the attention of the monkey or to threaten the animal.
3. Tourists are forbidden from leaving behind or offering any food or drinks to the monkeys. Tourists are strictly prohibited from throwing away or leaving behind any rubbish.
4. Tourists are not allowed to alter or to take any forest products or animals, or any other materials that are part of the natural habitats of the monkey.
5. Tourists are prohibited from smoking in the forest when observing monkeys.
6. Tourists must wear their life jackets at all times while on the boat. Tourists are prohibited from disembarking while in the forest, or making any attempt to do so.
7. Tourists may take photographs of the monkeys for personal use using cameras, but the flash mode should be switched off. Please note that taking photographs or footage of the animals for commercial purposes is not allowed in Sabah unless with special permission from the relevant local authorities such as the Wildlife Department.
8. Tour operators should observe a quota and plan their tour frequency so as not to disturb or agitate the animals.
9. It is suggested that not more that three boats should congregate at any one point on the river to observe a proboscis monkey group or individual monkey.

10. A distance of at least 10 metres should be maintained between observers and the monkeys at all times when observing the monkeys in the forest.
11. Monkeys intending to cross the river or in the midst of crossing the river should be left alone.
12. Boat operators are encouraged to switch off their outboard motor engine immediately upon encountering a proboscis monkey group or an individual. The monkeys can then be approached by slowly paddling the boat towards the animals.
13. High-power outboard motor engines are not encouraged as such engines make a loud noise and may disturb or agitate the monkeys. Recommended engines are between 15 and 50 hp only.

Community based ecotourism is increasingly becoming an important economic activity for the local community.

View of the forest at sunrise.

PLACES TO VISIT

Klias Peninsula: Kampong Garama, Kota Klias and Weston Wetlands

The nearest region to the main city of Kota Kinabalu where proboscis monkeys can be found is the Klias Peninsula, located about 90 kilometres to the southwest. The proboscis monkey population in Klias is the third largest in Sabah and the only major population on the west coast. Much of the proboscis monkey population here is concentrated in Garama within the Padas Damit Forest Reserve, in the central part of the Klias Peninsula. Other main populations are found in Weston wetlands, in the south of the peninsula and Kota Klias, near Beaufort town. These locations have sound infrastructure and are served by professional tour operators as well as local community programmes.

Sandakan Labuk Bay

In the centre of the mangrove forests of Sumawang in Sandakan is the Labuk Bay Proboscis Monkey Sanctuary. This privately-owned sanctuary located within an oil palm estate allows the chance to observe the proboscis monkeys up close without having to go on a wild boat ride. Lodges are available for overnight stays at Labuk Bay.

Kinabatangan

The Kinabatangan River, 560 kilometres long, is the longest river in Sabah; the southern branch originates in the Kuamut Highlands, and the northern branch from the Trus Madi Range and Labuk Highlands in the southwest of Sabah to the Sulu Sea in the east. This area sustains one of the world's richest ecosystems and also the largest proboscis monkey population in Sabah. Much of the Kinabatangan proboscis monkey population lives in the Lower Kinabatangan region in the district of Sandakan within

Homestay in one of the idyllic houses in Sukau.

the Kinabatangan Wildlife Sanctuary, although major populations can be found from Kampong Abai within the estuaries of Kuala Kinabatangan through to Batu Puteh more than 100 kilometres from the coast. Populations are most dense in the vicinity of Kampong Sukau, including around the six-kilometre-long Menanggul River (a tributary of the Kinabatangan River), Tenegang Besar River, Rasang River and other tributaries near Kampong Sukau. Many major local tour companies run lodges in the Sukau area with packages that include accommodation, transportation, meals and guided tours. Homestay programmes available in Sukau and the surrounding villages provide good opportunities to truly experience the life of the River People.

Lower Segama is situated in Lahad Datu district in eastern Sabah at the lower reaches of the Segama River. Proboscis monkeys are found along the Segama River from Terusan Segama near the coast past Kampong Litang upstream. The regions near Kampong Tidung, in particular Tabin River and Kampong Litang, are most densely populated. The proboscis monkey population here is the second largest in Sabah. Infrastructure for tourism in the Lower Segama regions is still not well established and road access is poor. Few tour agencies offer package tours, but a homestay is available at Kampong Tidung.

The proboscis monkey, an important ecotourist attraction in Sabah.

Bibliography

Agoramoorthy, G., Alagappasamy, C. and Hsu, M.J. (2004). Can proboscis monkeys be successfully maintained in captivity? A case of swings and roundabouts. *Zoo Biology* 23: 533–544.

Bennett, E.L. (1986). Proboscis monkeys in Sarawak: Their ecology, status, conservation and management. WWF-Malaysia/NYZS, NTIS.

Bennett, E.L. (1988). Proboscis monkeys and their swamp forests in Sarawak. *Oryx* 22: 69–74.

Bennett, E.L. and Gombek, F. (1993). *Proboscis Monkeys of Borneo.* Natural History Publications (Borneo), Kota Kinabalu.

Bennett, E.L. and Sebastian, A.C. (1988). Social organisation and ecology of proboscis monkey (*Nasalis larvatus*) in mixed coastal forest in Sarawak. *International Journal of Primatology* 9: 233–255.

Bernard, H. (1997). Some aspects of behaviour and feeding ecology of the proboscis monkey, *Nasalis larvatus*, based on a brief study in the Klias Peninsula, Sabah. *Borneo Science* 3: 5–84.

Bernard, H. (2009). Conservation of the last remaining populations of proboscis monkey on the west coast of Sabah. A final report submitted to the Ministry of Higher Education Malaysia. Project No. FRG0085-BD-1/2006.

Bernard, H. and Zulhazman, H. (2006). Population size and distribution of the proboscis monkey (*Nasalis larvatus*) in the Klias Peninsula, Sabah, Malaysia. *Malayan Nature Journal* 59: 153–163.

Bernard, H., Said, I.M. and Sha, C.M. (2006). Proboscis monkey protection plan for Klias Forest Reserve and its surrounding southern ecotones. Report, Conservation and Sustainable Use of Tropical Peat Swamp Forests and Associated Wetland Ecosystems, United Nations Development Programme/Global Environment Facility (UNDP/GEF) funded project MAL/99/G31.

Bernard, H., Matsuda, I., Hanya, G. and Ahmad, A.H. (in press). Characteristics of proboscis monkey (*Nasalis larvatus*) night sleeping-trees in Sabah, Malaysia. *International Journal of Primatology.*

Bismark, M. (2010). Proboscis monkey (*Nasalis larvatus*): Bio-ecology and conservation. In: Gursky, S. and Supriatna, J. (eds.). *Indonesian Primates*. Springer, New York, pp. 217–233.

Boonratana, R. (1993). The ecology and behaviour of the proboscis monkey (*Nasalis larvatus*) in the Lower Kinabatangan, Sabah. PhD thesis, Faculty of Graduate Studies, Mahidol University, Thailand.

Boonratana, R. (2000). Ranging behavior of proboscis monkeys (*Nasalis larvatus*) in the lower Kinabatangan, Northern Borneo. *International Journal of Primatology* 21: 497–518.

Boonratana, R. (2002). Feeding ecology of proboscis monkeys (*Nasalis larvatus*) in the Lower Kinabatangan, Sabah, Malaysia. *Sabah Parks Nature Journal* 6: 1–26.

Borneo Post (2006). River bank encroachers to face stern action: CM. 4 May.

Brandon-Jones, D. (1984). Colobus and leaf monkeys. In: Macdonald, D. (ed.). *The Encyclopaedia of Mammals*. London: Allen and Unwin, Vol. 1 pp. 398–408.

Brandon-Jones, D. (1996). The Asian colobinae (Mammalia: Cercopithecidae) as indicators of Quaternary climatic change. *Biological Journal of the Linnean Society* 59: 327–350.

Brandon-Jones, D. (1998). Pre-glacial Bornean primate impoverishment and Wallace's line. In: Hall, R. and Holloway, J.D. (eds.). *Biogeography and Geological Evolution of SE Asia*. Backbuys Publishers, Leiden (Netherlands).

Caldecott, J. (1992). Hunting patterns and their significance in Sarawak. In: Ismail, G., Mohamed, M. and Omar, S. (eds.). *Forest Biology and Conservation in Borneo*. Yayasan Sabah Center for Borneo Studies, Publication No. 2, Kota Kinabalu, Sabah, pp. 245–260.

Cheney, D.L. and Wrangham, R.W. (1987). Predation. In: Smuts, B.B., Cheney, D.L., Seyfarth, R.M., Wrangham, R.W. and Struhsaker, T.T. (eds.). *Primate Societies*. University of Chicago Press, Chicago, pp. 227–239.

Cleary, M.C. (1992). Plantation agriculture and the formulation of native land rights in British North Borneo, 1989–1930. *Geographical Journal* 158: 170–181.

Collins, L. and Roberts, M. (1978). Arboreal folivores in captivity: Maintenance of a delicate minority. In: Montgomery, G.G. (ed.). *The Ecology of Arboreal Folivores*. Smithsonian Institute Press, Washington, pp. 5–12.

Collins, N.M., Sayer, J.A. and Whitmore, T.C. (1991). *The Conservation Atlas of Tropical Forests, Asia and the Pacific*. IUCN, Macmillan, London.

Cranbrook, G.G. (1977). *Mammals of Borneo: Field Keys and an Annotated Checklist*. MBRAS Monograph No. 7, Malaysian Branch of the Royal Asiatic Society, Kuala Lumpur.

Davies, A.G. (1991). Seed-eating by red leaf monkeys (*Presbytis rubicunda*) in the dipterocarp forest of northern Borneo. *International Journal of Primatology* 12: 119–144.

Davies, A.G. and Baillie, I.C. (1988). Soil-eating by red leaf monkeys (*Presbytis rubicunda*) in Sabah, northern Borneo. *Biotropica* 20: 252–258.

Davies, A.G. and Oates, J. (1994). *Colobine Monkeys: Their Ecology, Behaviour and Evolution*. Cambridge University Press, Cambridge.

Davies, A.G. and Payne, J. (1982). A faunal survey of Sabah. IUCN/WWF Project No. 1692, WWF-Malaysia, Kuala Lumpur.

Davies, A.G., Caldecott, J.O. and Chivers, D.J. (1983). Natural foods as a guide to the nutrition of old world primates. In: Remfry, J. (ed.). *Standards in Laboratory Animal Management*. UFAW, Potters Bar, pp. 225–244.

Davies, A.G., Bennett, E.L. and Waterman, P.G. (1988). Food selection by two Southeast Asian colobine monkeys (*Presbytis rubicunda* and *Presbytis melalophos*) in relation to plant chemistry. *Biological Journal of the Linnean Society* 34: 33–56.

Davis, D.D. (1962). Mammals of the lowland rainforests of North Borneo. *Bulletin of the Singapore Natural History Museum, Singapore* 31: 1–129.

De Souza, L.L., Ferrari, S.F., Costa, M.L. and Kern, D.C. (2002). Geophagy as a correlate of folivory in red-handed howler monkeys (*Alouatta belzebul*) from Eastern Brazilian Amazonia. *Journal of Chemical Ecology* 28: 1613–1621.

Dela, J.D.S. (2007). Seasonal food use strategies of *Semnopithecus vetulus nestor*, at Panadura and Piliyandala, Sri Lanka. *International Journal of Primatology* 28: 607–626.

Di Fiore, A. (2003). Ranging behavior and foraging ecology of lowland woolly monkeys (*Lagothrix lagotricha poeppigii*) in Yasuní National Park, Ecuador. *American Journal of Primatology* 59: 47–66.

Dierenfeld, E.S., Koontz, F.W. and Goldstein, R.S. (1992). Feed intake, digestion and passage of the proboscis monkey (*Nasalis larvatus*) in captivity. *Primates* 33: 399–405.

Disotell, T.R. (2003). Primates: phylogenetics. In: *Encyclopedia of the Human Genome*, Nature Publishing Group, London.

Fashing, P.J. (2001). Activity and ranging patterns of guerezas in the Kahamega forest: intergroup variation and implications for intragroup feeding competition. *International Journal of Primatology* 22: 549–577.

Fashing, P.J. (2001). Feeding ecology of guerezas in the Kakamega Forest, Kenya: the importance of Moraceae fruit in their diet. *International Journal of Primatology* 22: 579–609.

Fashing, P.J. (2007). African colobine monkeys: patterns of between-group interaction. In: Campbell, C.J., Fuentes, A., Mackinnon, K.C., Panger, M. and Bearder, S.K. (eds.). *Primates in Perspective*. Oxford University Press, Oxford, pp. 201–224.

Fischer, J. and Lindenmayer, D.B. (2000). An assessment of the published results of animal relocations. *Biological Conservation* 96: 1–11.

Food and Agriculture Organisation (2002). An overview of forest products statistics in South and Southeast Asia. EC-FAO Partnership Programme (2000–2002) Tropical Forestry Budget Line, B7-6201/1B/98/0531. Project No. GCP/RAS/173/EC.

Fox, J.E.D. (1978). The natural vegetation of Sabah, Malaysia: The physical environment and classification. *Tropical Ecology* 19: 218–239.

Galdikas, B.M.F. (1985). Crocodile predation on a proboscis monkey in Borneo. *Primates* 26: 495–496.

Goossens, B., Setchell, J.M., Abulani, D.M.A., Jalil, F., James, S.S., Aris, S.H., Lakim, M.H., Seventri, A.D., Sariningsih, S.S. and Ancrenaz, M. (2002). A boat survey of primates in the Lower Kinabatangan Wildlife Sanctuary. In: Maryati, M., Takano, A.B., Goossens, B. and Indran, R. (eds). *Lower Kinabatangan Scientific Expedition.* Universiti Malaysia Sabah, Kota Kinabalu, pp. 37–45.

Gravitol, A.D., Ballou, J.D. and Fleischer, R.C. (2001). Microsatellite variation within and among recently fragmented populations of the golden lion tamarin (*Leopithecus rosalia*). *Conservation Genetics* 2: 1–9.

Groves, C.P. (1970). The forgotten leaf-eaters, and the phylogeny of the Colobinae. In: Napier, J.R. and Napier, P.H. (eds). *Old World Monkeys: Evolution, Systematics and Behavior.* Academic Press, New York, pp. 555–587.

Groves, C.P. (2001). *Primate Taxonomy.* Smithsonian Press, Washington, DC.

Guha, B. and Sha, C.M. (2010). Proboscis monkeys on Borneo: Who "nose" what the future holds? In: Dick, G. and Gusset, M. (eds). *Building a Future for Wildlife: Zoos and Aquariums Committed to Biodiversity Conservation,* World Association of Zoos and Aquariums (WAZA), Gland, Switzerland.

IUCN/SSC Conservation Breeding Specialist Group (2004). Proboscis Monkey Population Habitat Viability Analysis (PHVA), 2–7 December 2004, Bogor, Indonesia.

Jeffrey, S.M. (1982). Threats to the proboscis monkeys. *Oryx* 16: 337–339.

John, D.W. (1974). The timber industry and forest administration in Sabah under Chartered Company rule. *Journal of Southeast Asian Studies* 5: 55–81.

Kawabe, M. and Mano, T. (1972). Ecology and behaviour of the wild proboscis monkey, *Nasalis larvatus* (Wurmb) in Sabah, Malaysia. *Primates* 13: 213–228.

Kern, J.A. (1963). Observations on the habits of the proboscis monkey, *Nasalis larvatus* (Wurmb), made in the Brunei Bay area, Borneo. *Zoologica* 49: 183–191.

Kilbourn, A.M., Bosi, E.J., Wolfe, N.D., Andau, M. and Karesh, W.B. (2000). Disease evaluation and preventive medicine programs incorporated into ape conservation in Sabah, Malaysia. In: *The Apes: Challenges for the 21st Century*, Brookfield Zoo, May 10–13, 2000, Conference Proceedings, Chicago Zoological Society, Brookfield, Illinois, p. 376.

Kirkpatrick, R.C. (2007). The Asian colobines: Diversity among leaf-eating monkeys. In: Campbell, C.J., Fuentes, A., Mackinnon, K.C., Panger, M. and Bearder, S.K. (eds.). *Primates in Perspective*. Oxford University Press, Oxford, pp. 186–200.

Koenig, A. (2000). Competitive regimes in forest-dwelling Hanuman leaf monkey females (*Semnopithecus entellus*). *Behavioral Ecology and Sociobiology* 48: 93–109.

Kremen, C., Merenlender, A.M. and Murphy, D.D. (1994). Ecological monitoring: A vital need for integrated conservation and development programs in the tropics. *Conservation Biology* 8: 388–397.

Lambert, J.E. (2007). Primate nutritional ecology: Feeding biology and diet at ecological and evolutional scales. In: Campbell, C.J., Fuentes, A., Mackinnon, K.C., Panger, M. and Bearder, S.K. (eds.). *Primates in Perspective*. Oxford University Press, Oxford, pp. 482–495.

Lawrence, J., McCann, M.S. and Dierenfeld, E.S. (2005). Chemical composition of foods eaten by African colobines compared with Southeast Asian colobines. In: Proceedings Nutrition Advisory Group, Omaha.

Low, H. (1990). *Sarawak, Its Inhabitants and Productions: Being Notes during a Residence in that Country with his Excellency Mr. Brooke*. Reprint. Asian Heritage Series, Malaysia Delta, Petaling Jaya, Selangor.

Mackinnon, K. (1987). Conservation status of primates in Malaysia, with special reference to Indonesia. *Primate Conservation* 8: 175–183.

Maisels, A., Gautier-Hion, A. and Gautier, J.P. (1994). Diets of two sympatric colobines in Zaire: More evidence on seed-eating in forests on poor soils. *International Journal of Primatology* 15: 681–701.

Malim, T.P., Andau, M. and Ambu, L. (1999). A faunal survey of the Kalabakan Forest Reserve of Tawau, Sabah and its potential management implications. Sabah Wildlife Department report.

Manansang, J., Traylor-Holzer, K., Reed, D. and Leus, K. (2005). Indonesian proboscis monkey population and habitat viability assessment: final report. IUCN/ SSC Conservation Breeding Specialist Group, Apple Valley, MN.

Marsh, C.W. (1995). Danum Valley Conservation Area, Sabah, Malaysia management plan 1995–2000. Yayasan Sabah/Innoprise Corporation Sdn. Bhd., Kota Kinabalu.

Marsh, C.W. and Wilson, W.L. (1981). Effects of natural habitat differences and disturbances on the abundance of Malaysian primates. *Malaysian Journal of Applied Biology* 10: 227–249.

Matsuda, I. (2008). Feeding and ranging behaviors of proboscis monkey *Nasalis larvatus* in Sabah, Malaysia. PhD thesis, Graduate School of Environmental Earth Science, Hokkaido University, Japan.

Matsuda, I., Tuuga, A., Akiyama, Y. and Higashi, S, (2008), Selection of river crossing location and sleeping site by proboscis monkeys (*Nasalis larvatus*) in Sabah, Malaysia. *American Journal of Primatology* 70: 1097–1101.

Matsuda, I., Tuuga, A. and Higashi, S. (2008). Clouded leopard (*Neofelis diardi*) predation on proboscis monkeys (*Nasalis larvatus*) in Sabah, Malaysia. *Primates* 49: 227–231.

Matsuda, I., Tuuga, A. and Higashi, S. (2009). The feeding ecology and activity budget of proboscis monkeys. *American Journal of Primatology* 71: 478–492.

Matsuda, I., Tuuga, A. and Higashi, S. (2009). Ranging behaviour of proboscis monkeys in a riverine forest with special reference to ranging in inland forest. *International Journal of Primatology* 30: 313–325.

Matsuda, I., Kubo, T., Tuuga, A. and Higashi, S. (2010). A Bayesian analysis of the temporal change of local density of proboscis monkeys: Implications for environmental effects on a multilevel society. *American Journal of Physical Anthropology* 142: 235–245.

Matsuda, I., Tuuga, A. and Higashi, S. (2010). Effects of water level on sleeping-site selection and inter-group association in proboscis monkeys: Why do they sleep alone inland on flooded days? *Ecological Research* 25: 475–482.

Matsuda, I., Tuuga, A. and Bernard, H. (in press). Riverine refuging by proboscis monkeys (*Nasalis larvatus*) and sympatric primates: Implications for adaptive benefits of the riverine habitat. *Mammalian Biology.*

Mbora, D.N.M. and Meikle, D.B. (2004). Forest fragmentation and the distribution, abundance and conservation of the Tana River red colobus (*Procolobus rufomitratus*). *Biological Conservation* 118: 67–77.

McKey, D. and Waterman, P.G. (1982). Ranging behaviour of a group of black colobus (*Colobus satanas*) in the Douala-Edea Reserve, Cameroon. *Folia Primatologica* 39: 264–304.

Meijaard, E. and Nijman, V. (1999). The local extinction of the proboscis monkey *Nasalis larvatus* in Pulau Kaget Nature Reserve, Indonesia. *Oryx* 34: 66–70.

Meijaard, E. and Nijman, V. (2000). Distribution and conservation of the proboscis monkey (*Nasalis larvatus*) in Kalimantan, Indonesia. *Biological Conservation* 92: 15–24.

Miller, L.E. and Treves, A. (2007). Predation on primates. In: Campbell, C.J., Fuentes, A., Mackinnon, K.C., Panger, M. and Bearder, S.K. (eds.). *Primates in Perspective.* Oxford University Press, Oxford, pp. 525–543.

Mills, L.S. and Smouse, P.E. (1994). Demographic consequences of inbreeding in remnant populations. *The American Naturalist* 144: 412–431.

Mittermeier, R.A. and Cheney, D.L. (1987). Conservation of primates and their habitats. In: Smuts, B.B., Cheney, D.L., Seyfarth, R.M., Wrangham, R.W. and Struhsaker, T.T. (eds.). *Primate Societies.* University of Chicago Press, Chicago, pp. 477–490.

Murai, T. (2004). Social structure and mating behavior of proboscis monkey *Nasalis larvatus* (Primates; Colobinae). PhD thesis, Graduate School of Environmental Earth Science, Hokkaido University, Japan.

Murai, T. (2004). Social behaviors of all-male proboscis monkeys when joined by females. *Ecological Research* 19: 451–454.

Murai, T. (2006). Mating behaviors of the proboscis monkey (*Nasalis larvatus*). *American Journal of Primatology* 68: 832–837.

Myers, N., Mittermeier, R.A., Mittermeier, C.G., Da Fonseca, G.A.B. and Kent, J. (2000). Biodiversity hotspots for conservation priorities. *Nature* 403: 853–858.

Napier, J.R. and Napier, P.H. (1967). *A Handbook of Living Primates.* Academic Press, London.

Napier, P.H. (1985). *Catalogue of Primates in the British Museum (Natural History) and Elsewhere in the British Isles. Part III. Family Cercopithecidae, Subfamily Colobinae.* British Museum (Natural History), London.

New Sabah Times (2003). Proboscis at risk: Call for urgency as abandoned baby becomes first victim. 23 January.

Newton, P.N. (1992). Feeding and ranging patterns of forest hanuman leaf monkeys (*Presbytis entellus*). *International Journal of Primatology* 13: 245–285.

Noë, R. and Bshary, R. (1997). The formation of red colobus-diana monkey associations under predation pressure from chimpanzees. *Proceedings of the Royal Society of London B* 264: 253–259.

Nowell, K. and Jackson, P. (1996). Wild cats: Status survey and conservation action plan. IUCN, Gland, pp. 66–69.

Oates, J.F. (1987). Food distribution and foraging behavior. In: Smuts, B.B., Cheney, D.L., Seyfarth, R.M., Wrangham, R.W. and Struhsaker, T.T. (eds.). *Primate Societies.* University of Chicago Press, Chicago, pp. 197–209.

Olsen, D.M. and Dinerstein, E. (2002). The Global 200: Priority ecoregions for global conservation. *Annals of the Missouri Botanical Garden* 89: 199–224.

Onuma, M. (2002). Daily ranging patterns of the proboscis monkey (*Nasalis larvatus*) in coastal areas of Sarawak, Malaysia. *Mammal Study* 27: 141–144.

Partners of Community Organisations (PACOS). Indigenous peoples of Sabah. Accessed at http://www.sabah.net.my/PACOS/people.htm.

Payne, J. (1988). Orang-utan conservation in Sabah. WWF Malaysia Project No. 96/86. WWF International Project No. 3759.

Pinto, L.P. and Setz, E.Z.F. (2004). Diet of *Alouatta belzebul discolor* in an Amazonian rain forest of northern Mato Grosso State, Brazil. *International Journal of Primatology* 25: 1197–1211.

Pope, T.R. (1996). Socioecology, population fragmentation, and patterns of genetic loss in endangered primates. In: Avise, J., Hamrick, J. (eds.). *Conservation Genetics: Case Histories from Nature.* Kluwer Academic Publishers, Norwell, MA, pp. 119–159.

Primack, R.B. and Hall, P. (1992). Biodiversity and forest change in Malaysian Borneo. *BioScience* 42: 829–837.

Raemaekers, J.J. and Chivers, D.J. (1980). Socio-ecology of Malayan forest primates. In: Chivers, D.J. (ed.). *Malayan Forest Primates: Ten Years' Study in Tropical Rain Forest.* Plenum, New York, pp. 279–316.

Rajananthan, R. (1991). Differential habitat use by primates in Samunsam Wildlife Sanctuary, Sarawak, and its application to conservation management. MSc dissertation, University of Florida.

Rajananthan, R. (1995). A mammal and bird survey in the Lower Segama Region, Sabah with emphasis on the proboscis monkey. Sabah Wildlife Department and World Wildlife Fund, Malaysia.

Rajananthan, R. and Bennett, E.L. (1990). Notes on the social behaviour of wild proboscis monkey (*Nasalis larvatus*). *Malayan Nature Journal* 44: 35–44.

Ross, C. (1992). Basal metabolic rate, body weight, and diet in primates: an evaluation of the evidence. *Folia Primatologica* 58: 7–23.

Rowe, N. (1996). *The Pictorial Guide to the Living Primates.* Pogonias Press, Charlestown, RI.

Sabah Forestry Department (1989). *Forestry in Sabah.* Sandakan.

Sabah State Government (1998). *Forestry in Sabah — Status, Policy and Actions.* Government Printer, Sabah.

Sabah Forestry Department (2002). *Production and Export Statistics of Forest Products, 2001.*

Sabah Wildlife Department (2003). Faunal survey. Sabah Wildlife Department report.

Salter, R.E. and Mackenzie, N.A. (1985). Conservation status of the proboscis monkey in Sarawak. *Biological Conservation* 33: 119–132.

Salter, R.E., Mackenkie, N.A., Aken, K.M. and Chai, P.K. (1985). Habitat use, ranging behaviour, food habits of the proboscis monkey, *Nasalis larvatus* (van Wurmb), in Sarawak. *Primates* 26: 436–451.

Schultz, A.H. (1942). Growth and development of the proboscis monkey. *Bulletin of the Museum of Comparative Zoology, Harvard College* 89: 279–314.

Schultz, C.H. and Beck, J. (1999). A record of proboscis monkey (*Nasalis larvatus*) (Mammalia, Primates, Cercopithecidae) from Kinabalu Park, Sabah, Malaysia. *Sabah Parks Nature Journal* 2: 23–26.

Sebastian, A.C. (1994). A preliminary investigation of the proboscis monkey population in Danau Sentarum Wildlife Reserve, Western Kalimantan, Indonesia. Directorate General for Forest Protection and Nature Conservation (PHPA) and Asian Wetland Bureau, Bogor.

Sha, C.M. (2006). Distribution, abundance and conservation of proboscis monkeys in Sabah, Malaysia. Masters dissertation, Universiti Malaysia Sabah.

Sha, C.M. and Bernard, H. (2006). Proboscis odyssey 2006. *Malaysian Naturalist* 60: 9–13.

Sha, C.M., Bernard, H. and Nathan, S. (2008). Status and conservation of proboscis monkeys (*Nasalis larvatus*) in Sabah, East Malaysia. *Primate Conservation* 23: 107–120.

Soendjoto, M.A. (2004). A new record on habitat of the proboscis monkey (*Nasalis larvatus*) and its problems in south Kalimantan, Indonesia. *Tigerpaper* 31(2): 17–18.

Srikwan, S. and Woodruff, D. (2000). Genetic erosion in isolated small mammal populations following rainforest fragmentation. In: Young, A.G. and Clarke, G.M. (eds.). *Genetics, Demography and Viability of Fragmented Populations*. Cambridge University Press, New York, pp. 149–172.

Stanford, C.B. (1991). *The Capped Leaf Monkey in Bangladesh: Behavioral Ecology and Reproductive Tactics*. Karger, Basel.

Stanford, C.B. (1991). The diet of the capped Leaf monkey (*Presbytis pileata*) in a moist deciduous forest in Bangladesh. *International Journal of Primatology* 12: 199–216.

Sterk, E.H.M. (1997). Determinants of female dispersal in Thomas leaf monkeys: Are they linked to human disturbance? *American Journal of Primatology* 42: 49–198.

Stibig, H.-J. and Malingreau, J.P. (2003). Forest cover of insular Southeast Asia mapped from recent satellite images of coarse spatial resolution. *Ambio* 32: 469–475.

Struhsaker, T.T. and Leland, L. (1979). Socioecology of five sympatric monkey species in the Kibale Forest, Uganda. In: Rosenblatt, J., Hinde, R., Beer, C. and Busnel, M. (eds.). Advances in the Study of Behavior, Vol. 9, Academic Press, New York, pp. 159–228.

Teichroeb, J.A., Saj, T.L., Paterson, J.D. and Sicotte, P. (2003). Effect of group size on activity budgets of *Colobus vellerosus* in Ghana. *International Journal of Primatology* 24: 743–758.

Terborgh, J. (1983). *Five New World Primates: A Study in Comparative Ecology.* Princeton University Press.

Whitmore, T.C. (1984). *Tropical Rain Forests of the Far East*, 2nd edn. Clarendon, Oxford.

Wilson, C.C. and Wilson, W.L. (1975). Methods for censusing forest-dwelling primates. In: Kondo, S., Kawai, M. and Ehara, A. (eds.). *Contemporary Primatology.* Karger, Basel, pp. 345–350.

World Wildlife Fund-Germany (2005). *Borneo: Treasure Island at Risk.* WWF-Germany, Frankfurt.

World Wildlife Fund-Malaysia (1992). Sabah conservation strategy Vol. 1: Background and analysis. WWF-Malaysia and the Ministry of Tourism and Environmental Development, Sabah, Malaysia.

World Wide Fund-Malaysia (1992). Sabah conservation strategy — Final Report. World Wide Fund for Nature, Malaysia, Kota Kinabalu,

World Wide Fund-Malaysia (1997). The Kinabatangan floodplain.

Wrangham, R.W. (1980). An ecological model of female-bonded primate groups. *Behavior* 75: 262–300.

Yeager, C.P. (1989). Feeding ecology of the proboscis monkey. *International Journal of Primatology* 10: 497–530.

Yeager, C.P. (1990). Proboscis monkey (*Nasalis larvatus*) social organization: Group structure. *American Journal of Primatology* 20: 95–106.

Yeager, C.P. (1990). Notes on the sexual behaviour of the proboscis monkey. *American Journal of Primatology* 21: 223–227.

Yeager, C.P. (1991). Possible antipredator behaviour associated with river crossings by proboscis monkeys (*Nasalis larvatus*). *American Journal of Primatology* 24: 61–66.

Yeager, C.P. (1991). Proboscis monkey (*Nasalis larvatus*) social organization: Intergroup patterns of association. *American Journal of Primatology* 23: 73–86.

Yeager, C.P. (1992). Changes in proboscis monkey (*Nasalis larvatus*) group size and density at Tanjung Puting National Park, Kalimantan Tengah, Indonesia. *Tropical Biodiversity* 1: 49–55.

Yeager, C.P. (1992). Proboscis monkey (*Nasalis larvatus*) social organization: Nature and possible functions of intergroup patterns of association. *American Journal of Primatology* 26: 133–137.

Yeager, C.P. (1993). Ecological constraints on the intergroup associations in the proboscis monkey *N. larvatus*. *Tropical Biodiversity* 1: 89–100.

Yeager, C.P. (1995). Does intraspecific variation in social systems explain reported differences in the social structure of the proboscis monkey (*Nasalis larvatus*)? *Primates* 36: 575–582.

Yeager, C.P. and Blondal, T.K. (1992). Conservation status of the proboscis monkeys (*Nasalis larvatus*) at Tanjung Puting National Park, Kalimantan Tengah, Indonesia. In: Ismail, G., Mohamed, M. and Omar, S. (eds.). *Forest Biology and Conservation in Borneo*. Yayasan Sabah Centre for Borneo Studies, Publication No. 2. Kota Kinabalu, Sabah, pp. 220–228.

Yeager, C.P. and Frederiksson, G. (1998). Fire impacts on primates and other wildlife in Kalimantan, Indonesia during 1997/1998. WWF Indonesia, Jakarta.

Yeager, C.P., Silver, S.C. and Dierenfeld, E.S. (1997). Mineral and phytochemical influences on foliage selection by the proboscis monkey (*Nasalis larvatus*). *American Journal of Primatology* 41: 117–128.

Zhou, Q., Wei, F., Huang, C., Li, M., Ren, B. and Luo, B. (2007). Seasonal variation in the activity patterns and time budgets of *Trachypithecus francoisi* in the Nonggang Nature Reserve, China. *International Journal of Primatology* 28: 657–671.

Zoological Society of London (1976). *International Zoo Yearbook* 16.

Zoological Society of London (1998). *International Zoo Yearbook* 36.

Glossary

Aggregation: a group of organisms of the same or different species living closely together but less integrated than a society.

Agro-ecosystem: Sustainable use of resources within the stability and resilience of natural ecological systems.

Alcatraz: An analogical reference to Alcatraz island and its infamous high security military prisons.

Arboreal: living in trees or adapted for living in trees.

Bacteria: Extremely small, relatively simple prokaryotic microorganisms. Some cause disease in humans and domestic animals but some bacteria are useful in nature, for example bacteria that improve soil fertility and bacteria in cows, goats and colobine monkeys which help digest food.

Before Present: a time scale used in archaeology, geology, and other scientific disciplines to specify when events in the past occurred. Because the "present" time changes, standard practice is to use CE 1950 as the arbitrary origin of the age scale. For example, 1500 BP means 1500 years before 1950, that is, in the year 450 CE.

Biological diversity: the variation of life forms within a given ecosystem, biome, or for the entire Earth. Biodiversity is often used as a measure of the health of biological systems. The biodiversity found on Earth today consists of many millions of distinct biological species, which is the product of nearly 3.5 billion years of evolution.

Cercopithecidae or Old World monkeys: a primate family including 18 genera and 81 species native to Africa and Asia today. Cercopithecids are divided into two ecologically and morphologically distinct subfamilies. The cercopithecines are omnivorous, have cheek pouches, and simple stomachs; while the colobines are

Colobinae: a subfamily of Cercopithecidae that includes 58 species in 10 genera. Colobines are medium-sized primates with long tails and diverse colorations. Most species are arboreal, although some live a more terrestrial life. They are almost exclusively folivorous.

Delta: a delta is a landform that is created at the mouth of a river where that river flows into an ocean, sea, estuary, lake, reservoir, flat arid area, or another river. Deltas are formed from the deposition of the sediment carried by the river as the flow leaves the mouth of the river. Over long periods of time, this deposition builds the characteristic geographic pattern of a river delta.

Endangered: Under the IUCN Red List of Threatened Species, "Endangered" species are considered to be facing a very high risk of extinction in the wild.

Enzymatic digestion: The process by which food is converted into substances that can be absorbed and assimilated by the body. It is accomplished in the alimentary canal by the mechanical and enzymatic breakdown of foods into simpler chemical compounds.

Estuary: a semi-enclosed coastal body of water with one or more rivers or streams flowing into it, and with a free connection to the open sea.[1] It is affected by both marine influences, such as tides, waves, and the influx of saline water; and riverine influences, such as flows of fresh water and sediment. As a result it may contain many biological niches within a small area, and so is associated with high biological diversity.

Floodplain: flat or nearly flat land adjacent to a stream or river that experiences occasional or periodic flooding. It includes the floodway, which consists of the stream channel and adjacent areas that carry flood flows, and the flood fringe, which are areas covered by the flood, but which do not experience a strong current.

Fragmentation: The breaking up of an organism's habitat into discontinuous chunks, particularly for organisms that have difficulty moving from one of those chunks to another.

Girth at breast height: the circumference of a tree stem measured at about chest height from the ground.

Homestay: a private home which offers accommodation to paying visitors, providing a unique opportunity to experience the way of life of local people of an area along with indigenous and traditional cultures.

Indigenous: Any ethnic group of people who inhabit a geographic region with which they have the earliest known historical connection, alongside more recent immigrants who have populated the region and may be greater in number.

Interglacial: a geological interval of warmer global average temperature that separates glacial periods within an ice age. The current Holocene interglacial has persisted since the Pleistocene, about 11,400 years ago.

Kampong or Kampung (Malay): village.

Old World monkeys: See Cercopithecidae.

Omnivorous: Species that eat both plants and animals as their primary food source. They are opportunistic, general feeders not specifically adapted to eat and digest either meat or plant material exclusively.

Physiological adaptation: A metabolic or physiologic adjustment within the cell, or tissues, of an organism in response to an environmental stimulus resulting in the improved ability of that organism to cope with its changing environment.

Sacculated: having or formed of a series of saccular expansions.

Sedentary: pertaining to animals that move about little or are non-migratory.

Spatial heterogeneity: a property generally ascribed to a landscape or to a population. It refers to the uneven distribution of various concentrations of each species within an area. A population showing spatial heterogeneity is one where various concentrations of individuals of this species are unevenly distributed across an area; nearly synonymous with "patchily distributed".

Terrestrial: an animal that lives on the land, as opposed to in water, air, or in the trees.

Toxins: poisonous substances produced by living cells or organisms. Toxins can be small molecules, peptides, or proteins that are capable of causing disease on contact with or absorption by body tissues interacting with biological macromolecules such as enzymes or cellular receptors.

Acknowledgements

The inspiration for this book arose from a culmination of the authors' field research on proboscis monkeys in Sabah, East Malaysia. Over the years, our work on proboscis monkeys would not have been possible without the support of various institutions. We gratefully acknowledge funding support from: the Wildlife Reserves Singapore's Wildlife Conservation Fund and San Diego's Conservation and Research for Endangered Species (CRES) for the state-wide survey of proboscis monkeys in Sabah; the Fundamental Research Grant Schemes of the Ministry of Higher Education Malaysia (FRG0085-BD-1/2006), the Pro Natura Foundation Japan and Nature Conservation Society of Japan for research work in Klias Peninsula; and the JSPS core-to-core program HOPE, the Global COE Program (Establishment of Center for Integrated Field Environmental Science), MEXT, Japan, the Grant-in-Aid for Young Scientists (B) (project #21770261), and the JSPS International Training Program (ITP) for research work in Kinabatangan.

We are grateful to the individuals who have provided the opportunities and unwavering support for our work. From the Sabah Wildlife Department: the late Datuk Patrick Andau; Dr Laurentius Ambu, Augustine Tuuga, Dr Senthival Nathan and all the district officers and rangers; from Sabah Forestry Department: Datuk Sam Mannan, Hj. Hussin Tukiman, Ladwin Ruki, Dr John B. Sugau, Joan T. Pereira, Postar Miun; from the Economic Planning Unit: Munirah Abd. Manan, Gwendolen Vu; from Universiti Malaysia Sabah: Datuk Dr Mohd. Noh Dalimin, Datuk Seri Panglima Dr Kamaruzaman Hj. Ampon, Datin Dr Maryati Mohamed, Dr Abdul Hamid Ahmad; from HUTAN: Dr Isabelle Lackman-Ancrenaz, Dr Marc Ancrenaz, Zainal Abidin Jaafar and team; from Wildlife Reserves Singapore: Fanny Lai and Biswajit Guha; and from CRES: Dr Chia Tan and Dr Andy Phillips.

Numerous individuals have enriched our research perspectives through sharing their knowledge and advice, namely, Dr Ramesh Boonratana, Dr Erik Meijaard, Datuk Dr Junaidi Payne, Dr Menno Schilthuizen, Dr Seigo Higashi, Dr Juichi Yamagiwa, Dr Kunio Watanabe, Dr Naoki Agetsuma, Dr Toshio Iwakuma, Dr Takuya Kubo, Dr Yoshihiro Akiyama, Dr Goro Hanya, Dr Akisato Nishimura, Dr Tadahiro Murai, Dr Benoit Goossens, and Dr Konstans Wells.

Our most sincere gratitude goes to our local assistants and the hospitable orang kampung we were fortunate to meet during the course of our work in Sabah. Much of our work would not have been possible without their kind assistance and ensuring our safety in often precarious conditions. John would like to thank Ahmad, Badrul and Abidin. Ikki would like to thank Ahmad Bin Arsih, Mat Sarudin Bin Abd. Karim and Hartiman Bin Abd. Rahman. Henry would like to thank Awang Masis, Lee Shan Khee, Leong Ann Ying, Siti Zaraurah Ag. Gabor, Gilmoore Bolongan and Lucy Wong.

The authors would like to thank all the friends and colleagues for the critical discussions, insights and encouragement which helped sustain the conviction to push on over difficult times. We are also eternally indebted to our families for their understanding and moral support throughout the "vanished" years in the forests. Ikki would particularly like to thank Yuko Matsuda who has stood by him and helped him overcome numerous difficulties.

We wish to thank contributors to this publication: William Beavitt for the insights and photographic contributions from his work in Sarawak, Bjorn Olsen, Lester Ledesma and Dr Arthur Y.C. Chung for their invaluable photographic contributions.

Lastly, we wish to thank Datuk Chan Chew Lun of Natural History Publications (Borneo) for taking a keen interest in our project and for providing us with the opportunity to publish this book.

Photo Credits

Ch'ien C. Lee
ii, 2, 3, 5, 6, 8, 17, 18 (top left and top right),19 (top and bottom), 32, 34, 40, 44, 55 (all), 56, 58, 59, 62/63, 72, 74 (top and bottom), 76, 96

John Sha
viii, 11, 12 (top right), 13 (top right, bottom left and bottom right), 18 (bottom), 20, 22, 26, 29 (top right, bottom left and right), 30 (bottom), 49 (bottom), 61 (bottom), 65 (top and bottom), 69, 78 (Main and Inset), 80, 82, 83, 86, 88 (left and right), 90, 92, 105

Ikki Matsuda
12 (bottom right and bottom left), 13 (top left), 36, 45 (top left), 60 (all), 61 (top, center), 89, 102

Bjorn Olesen
21, 29 (top left), 30 (top), 45 (bottom), 46, 50, 51, 70/71, 84/85, 95, 104, 106

Lester Ledesma
24, 25 (left and right)

William Beavitt
45 (top right), 74 (bottom)

Arthur Chung
4, 28, 42, 43, 48, 49 (top), 52, 53, 98, 100, 101

Loi Pui King
10

Azrie Alliamat
12 (top-left)

Joseph Tangah
66

C.L. Chan
75

About the Authors

John Sha was born in Singapore in 1979. He graduated from the National University of Singapore with a B.Sc. in 2002 and joined the Singapore Zoo as Research and Conservation Officer. In 2004, he embarked on his research on proboscis monkeys in Sabah and completed his M.Sc. in Wildlife Conservation and Management with Universiti Malaysia Sabah in 2006. He subsequently worked with the United Nations Development Programme and the National Parks Board Singapore before rejoining the Wildlife Reserves Singapore in 2008 where he is currently Conservation and Research Curator. John's main research interest is in primates and his research approach is through basic applied science that is relevant to the holistic conservation of primates and other species both in the wild and in captivity.

Ikki Matsuda was born in Shizuoka, Japan in 1978. He has been studying the feeding ecology and ranging behaviour of wild proboscis monkeys in Kinabatangan since 2005 and obtained his Ph.D. from Hokkaido University in 2008. Ikki is currently a Research Fellow in the Department of Ecology and Social Behavior, Primate Research Institute, Kyoto University where he is continuing his research interest on proboscis monkeys and the environmental factors that influence the evolution of their social systems. He is also working towards a synthetic understanding of the evolution and ecology of colobine species, particularly the odd-nosed colobines.

Henry Bernard was born in Papar, Sabah in 1969. He received his B.Sc. (Honours) degree in Zoology from the Department of Zoology, University of Malaya, Kuala Lumpur in 1992. He obtained his M.Sc. degree in Wildlife Management and Control from the Department of Animal and Microbial Sciences, University of Reading, U.K. in 1996 and subsequently, Ph.D. in Zoology from the Vertebrate Department, Copenhagen University, Denmark in 2003. Henry worked as an Assistant Fisheries Officer and Wildlife Officer at the Fisheries Department and Wildlife Department of Sabah, respectively, from 1992 to 1995, before joining the Universiti Malaysia Sabah's Institute for Tropical Biology and Conservation as a lecturer in 1997. Henry started researching proboscis monkeys in the Klias Peninsula in 1994 and has continued his work since then. Henry's other interests include non-volant small mammal ecology, statistics in biology, protected area management and conservation biology in general.